QUALITY CRITERIA FOR WATER REUSE

Panel on Quality Criteria for Water Reuse
Board on Toxicology and Environmental Health Hazards
Commission on Life Sciences
National Research Council

NATIONAL ACADEMY PRESS
Washington, D. C. 1982

NOTICE: The project that is the subject of this report was approved by the Governing Board of the National Research Council, whose members are drawn from the Councils of the National Academy of Sciences, the National Academy of Engineering, and the Institute of Medicine. The members of the committee responsible for the report were chosen for their special competences and with regard for appropriate balance.

This report has been reviewed by a group other than the authors according to procedures approved by a Report Review Committee consisting of members of the National Academy of Sciences, the National Academy of Engineering, and the Institute of Medicine.

The National Research Council was established by the National Academy of Sciences in 1916 to associate the broad community of science and technology with the Academy's purposes of furthering knowledge and of advising the federal government. The Council operates in accordance with general policies determined by the Academy under the authority of its congressional charter of 1863, which establishes the Academy as a private, nonprofit, self-governing membership corporation. The Council has become the principal operating agency of both the National Academy of Sciences and the National Academy of Engineering in the conduct of their services to the government, the public, and the scientific and engineering communities. It is administered jointly by both Academies and the Institute of Medicine. The National Academy of Engineering and the Institute of Medicine were established in 1964 and 1970, respectively, under the charter of the National Academy of Sciences.

The study on which this report is based was requested and funded by the U.S. Department of Agriculture (Contract No. FSQS-14-W-80), the U.S. Environmental Protection Agency (Contract No. 68-01-3169), and the U.S. Army Corps of Engineers (Contract No. DACW 31-76-C-0069).

Library of Congress Catalog Card Number 82-61430

International Standard Book Number 0-309-03326-8

Available from

NATIONAL ACADEMY PRESS
2101 Constitution Avenue, N.W.
Washington, D.C. 20418

Printed in the United States of America

List of Participants

Panel on Quality Criteria for Water Reuse

RUSSELL F. CHRISTMAN, <u>Chairman</u>, Department of Environmental Sciences and Engineering, University of North Carolina, Chapel Hill, North Carolina
JULIAN ANDELMAN, University of Pittsburgh, Pittsburgh, Pennsylvania
JOSEPH C. ARCOS, Office of Toxic Substances, U.S. Environmental Protection Agency, Washington, D.C.
JOSEPH BORZELLECA, Health Sciences Division, Virginia Commonwealth University, Richmond, Virginia
THOMAS CLARKSON, Department of Radiation Biology, University of Rochester, Rochester, New York
ROSE DAGIRMANJIAN, Department of Pharmacology, University of Louisville, Louisville, Kentucky
RICHARD ENGELBRECHT, Department of Civil Engineering, University of Illinois, Urbana, Illinois
DAVID GAYLOR, Division of Biometry, National Center for Toxicological Research, Jefferson, Arkansas
HAROLD KALTER, Children's Hospital Research Foundation, Institute for Developmental Research, Cincinnati, Ohio
PERRY McCARTY, Department of Civil Engineering, Stanford University, Stanford, California
VERNE RAY, Pfizer, Inc., Groton, Connecticut
CHARLES RHODE, Department of Biostatistics, The Johns Hopkins University, Baltimore, Maryland

Consultants

IRWIN H. SUFFET, Drexel University, Philadelphia, Pennsylvania
ROBERT L. JOLLEY, Oak Ridge National Laboratory, Oak Ridge, Tennessee

National Research Council Staff

ROBERT J. GOLDEN, Project Director
ROBERT G. TARDIFF, Executive Director of the Board on Toxicology and Environmental Health Hazards
AGNES E. GASKIN, Secretary

Project Officers

BARTIE WOODS, U.S. Department of Agriculture, Washington, D.C.
LELAND J. McCABE, U.S. Environmental Protection Agency, Cincinnati, Ohio

Board on Toxicology and Environmental Health Hazards

RONALD W. ESTABROOK, Department of Biochemistry, University of Texas Medical School (Southwestern), Dallas, Texas, <u>Chairman</u>
PHILIP J. LANDRIGAN, National Institute for Occupational Safety and Health, Cincinnati, Ohio, <u>Vice-Chairman</u>
EDWARD BRESNICK, Department of Biochemistry, University of Vermont, Burlington, Vermont
THEORDORE L. CAIRNS, Greenville, Delaware (retired)
VICTOR H. COHN, Department of Pharmacology, George Washington University Medical Center, Washington, D.C.
A. MYRICK FREEMAN, Department of Economics, Bowdoin College, Brunswick, Maine
RONALD W. HART, National Center for Toxicological Research, Jefferson, Arkansas
MICHAEL W. LIEBERMAN, Department of Pathology, Washington University School of Medicine, St. Louis, Missouri
RICHARD A. MERRILL, School of Law, University of Virginia, Charlottesville, Virginia
ROBERT A. NEAL, Chemical Industry Institute of Toxicology, Research Triangle Park, North Carolina
IAN C. NISBET, Clement Associates, Washington, D.C.
JOHN M. PETERS, Department of Family and Preventive Medicine, University of Southern California, Los Angeles, California
LIANE B. RUSSELL, Biology Division, Oak Ridge National Laboratory, Oak Ridge, Tennessee
CHARLES R. SCHUSTER, JR., Department of Psychiatry, University of Chicago, Chicago, Illinois

Ex-Officio Members

JAMES F. CROW, Genetics Department, University of Wisconsin, Madison, Wisconsin
ROGER O. McCLELLAN, Lovelace Biomedical and Environmental Research Institute, Albuquerque, New Mexico
ROBERT E. MENZER, Department of Entomology, University of Maryland, College Park, Maryland
ROBERT W. MILLER, National Cancer Institute, Bethesda, Maryland
SHELDON D. MURPHY, Department of Pharmacology, University of Texas, Houston, Texas
NORTON NELSON, Institute of Environmental Medicine, New York University Medical Center, New York, New York
JAMES L. WHITTENBERGER, School of Public Health, Harvard University, Boston, Massachusetts

Contents

1 Introduction   1

2 Wastewater Reuse Systems   7

3 Chemical and Microbiological Constituents of Reuse Systems   17

4 Concentration Methods for Analysis and Toxicity Testing   51

5 Health Effects Testing   62

6 Strategies for Assessing and Monitoring Water Quality for Human Exposure   94

7 Assessment and Criteria for Potable Water Reuse   117

Appendixes

A Concentration Methodologies for Preparation of Water Concentrates for Toxicity Testing   135

B Further Statistical Details on Sampling   143

QUALITY CRITERIA FOR WATER REUSE

1
Introduction

Worldwide interest in water reuse for potable purposes has recently intensified; however, no criteria or standards currently exist to evaluate the quality of treated water produced from such "unacceptable" sources. Thus, the National Research Council's Committee to Review the Potomac Estuary Experimental Water Treatment Plant (U.S. Army Corps of Engineers project at Blue Plains, Washington, D.C.) charged the Panel on Quality Criteria for Water Reuse with advising the committee on standards of comparison for determining the suitability of water supplies produced from unacceptable sources such as wastewater. The committee functioned as a technical review body for the demonstration and pilot treatment plant project, which was designed and operated to evaluate the technical parameters involved in the advanced treatment of a blend of Potomac estuary water and secondary municipal wastewater treatment effluent. Although it asked the panel to provide specific guidance on health effects criteria for use in evaluating whether the effluent produced by the experimental treatment plant was suitable for human consumption, it was hoped that the panel's efforts would have a broader application in that the criteria would be appropriate for water reuse for drinking and food processing.

This report was also prepared for the U.S. Environmental Protection Agency (EPA) (charged with regulating the safety of drinking water) and the U.S. Department of Agriculture (USDA) (charged with the regulation of food processing) as a separate and more general statement of the areas of concern and of the panel's approach to the problem. The Food and Drug Administration (FDA) may also find this topic of interest in carrying out its regulatory responsibilities. The panel did not study water use and/reuse in the broad sense of the hydrologic cycle as affected by anthropogenic intervention. Nor did it analyze the current extent of, future need for, or desirability of categorical water reuse. The focus, rather, was on the scientific questions concerning the quality criteria that should be applied if water is to be reused in any of the senses implied in the following definitions.

REUSE DEFINITIONS

Reuse--Productive utilization of appropriately treated wastewater. Recycling is a special case of reuse wherein the wastewater originates with the user.

Direct Reuse--The piped connection of a wastewater effluent to the intake works of a water supply facility.

Indirect Reuse--The abstraction of water for productive use from a natural surface or underground source that is fed in part by discharge of wastewater effluent.

Potable Reuse--The direct or indirect utilization of wastewater effluent for potable purposes.

Nonpotable Reuse--The direct or indirect utilization of wastewater effluent for nonpotable purposes.

Tertiary Wastewater Treatment--Treatment step(s) beyond conventional secondary treatment for the purpose of increasing the percent removal of suspended solids and biological oxygen demand (BOD).

Advanced Wastewater Treatment (AWT)--Treatment step(s) beyond conventional secondary treatment generally including physicochemical methods for the removal of one or more of the following wastewater constituents: phosphorus, nitrogen, heavy metals, and synthetic organic chemicals.

REUSE AND TREATMENT

Wastewater reuse is a reasonable alternative for extending a water supply, and many reuse systems are operating today in areas where water is scarce. They provide water to irrigate agricultural lands, golf courses, and landscapes; to fill recreational lakes; to fulfill industrial needs; and to replenish groundwaters. No systems in use in the United States were planned to deliver reclaimed wastewater directly to the consumer for potable use.

Types of Reuse Systems

One issue that is the subject of considerable debate is direct versus indirect potable reuse. In direct reuse, reclaimed wastewater is immediately added to the drinking water supply of the community; in indirect reuse, reclaimed water is stored in a reservoir or allowed to percolate through the ground. The storage and, perhaps, dilution that indirect reuse provides between treatment and consumption allows time for natural events to purify the water further. Mixing with natural waters also helps reduce the concentration of contaminants. Planned indirect reuse systems in operation today include the Whittier Narrows and San Jose plants in Los Angeles County and Water Factory 21 in Orange County, Calif., where reclaimed waters are percolated or injected into groundwater supplies. The Manassas treatment plant in Virginia discharges directly into the Occoquan Reservoir, a primary drinking water supply. Such planned indirect reuse is perhaps similar to the unplanned indirect reuse that generally occurs when one city discharges its waste into a river or stream used by a downstream community for its water supply.

In proposing criteria for direct or indirect potable reuse, consideration should be given to whether the need is for a short-term emergency situation or for normal use over a prolonged period. However, regardless of the length of use, the major issue associated with reuse today concerns the chronic health problems that might result from ingesting the mixture of inorganic and organic materials that remains in water, even after subjecting it to the most advanced treatment methods. Potential chronic health risks might include cancer, birth defects, and genetic alterations; however, these would likely be of less importance for short-term use. For a longer period of use, the health effects would take on increased importance, and the aesthetic quality of the water and the possible presence of human pathogens would also be of concern, although the solutions to these problems are easier today than they were in recent decades. The much greater probability that adequately safe water could be provided for short-term emergencies than for long-term use should perhaps be considered when developing criteria for potable reuse.

Even though there are precedents that have been cited as evidence that indirect planned potable reuse has been accepted in some locations, there is inadequate information from which to judge the safety of such a practice. In the panel's opinion, U.S. drinking water regulations were not established to judge the suitability of raw water supplies heavily contaminated with municipal and industrial wastewaters. Thus, criteria to judge the relative safety of using heavily contaminated water supplies as part of the potable water supply--direct or indirect, planned or unplanned--need to be developed.

The Efficiency of Overall Reuse Systems

The effectiveness of a reuse system--to reduce the concentration of contaminants in a given source of water or to produce water with certain required characteristics--is a function of the operation of the overall water system, not just of the treatment portion of the system itself. For instance, wastewater can be segregated to exclude some of the more contaminated industrial wastes, or it can be taken for treatment at certain times of the day when contaminant levels are known to be reduced, thus providing a better quality source of water for reuse. The source wastewater can then be treated by normal primary and secondary methods, using a constant flow rate to increase efficiency and reliability. Following this step, the reclaimed water can be percolated through the soil or injected directly into an aquifer for storage. Passage through soils or aquifers provides additional treatment, and storage provides time for slower processes to effect removal. The reclaimed water can also be blended with natural surface or groundwater of higher quality, again to reduce the concentration of contaminants. In some instances, this combined water may undergo normal water treatment processes that provide an additional measure of contaminant removal and treatment reliability.

An additional factor is the sequence of reuse. If reclaimed wastewater is recycled and used again, certain constituents (such as dissolved salts) can increase to undesirable concentrations with each cycle. Thus, when evaluating the effectiveness of contaminant removal, the overall system must be considered, including the initial wastewater characteristics, the various blendings and treatments to which the water will be subjected, and the effect of storage on contaminant removal from reclaimed water.

QUESTIONS OF CONCERN

The panel recognized the existing water quality criteria for individual compounds and their relevance to the categorical reuse situation. Furthermore, it was aware of the quality criteria stated in the Safe Drinking Act (P.L. 93-523) and in various drinking water standards (e.g., those of the EPA and the World Health Organization). The panel also considered the growing scientific literature (National Academy of Sciences, 1977, 1980) indicating that most of the organic contaminants of domestic wastewater, industrial wastewater, publicly supplied drinking water, and natural surface waters are not reflected in these standards.

The panel also was aware of the present state of knowledge regarding the deficiency of toxicological data on compounds that have been identified in water. In the course of its deliberations, it was confronted with many important issues, including the following:

- the adequacy of present knowledge regarding organic and inorganic contaminants in wastewaters and water supplies;
- the effect of chemical isolation, concentration, and storage procedures on solute composition and concentration;
- the effects of changes in quality of treatment plant effluent with time and degree of treatment;
- the applicability of conventional toxicological testing procedures to the evaluation of complex and varying mixtures of chemical contaminants;
- the adequacy of procedures for developing drinking water standards from the results of toxicological testing;
- the adequacy of existing EPA drinking water standards, developed on the assumption that the highest quality source would be used;
- the suitability of organic compounds on EPA lists of priority pollutants as representative of those organic contaminants likely to be present in a water reuse situation;
- the suitability of surrogate chemical parameters for monitoring treated wastewater to indicate the possible presence of constituents harmful to human health;
- the utility of model monitoring and biotesting protocols for both the pilot-plant stage and normal operation of all wastewater reuse systems versus the need for *ad hoc* protocols for each such system;

- the inadequacy of current methods for the detection, enumeration, and identification of many important pathogenic microorganisms to permit thorough microbial characterization of contaminated waters and to determine their removal and/or inactivation by various treatment processes;
- the efficacy of the conventional indicator systems and/or the development of new systems to determine the microbial acceptability of reclaimed water for potable purposes;
- the availability of reliable real-time monitoring techniques to ensure the operational integrity of treatment systems by using infectious agents and/or indicators of treatment effectiveness and to provide information rapidly; and
- the feasibility of evaluating possible risks to human health from use of treated wastewater by comparison of its quality with that of currently used supplies of water from other sources.

APPROACH TO THE STUDY

Given the complex nature of the problems and the current interest in reuse of wastewater, the panel concentrated on preparing the best scientific statement possible regarding the nature and characteristics of specific reuse systems, the efficacy of current treatment technology, and the special monitoring needs created by specific reuse situations from a chemical and microbiological perspective. Finally, the panel attempted to offer the best practical scientific statement concerning health effects criteria for the evaluation of reused water intended for human consumption.

It is not possible to evaluate the health effects of the many compounds detected in the aquatic environment; thus, the panel adopted the view that the quality of reused water could be compared to that of conventional drinking water supplies, which are assumed to be safe. The philosophy behind the Interim Primary Drinking Water Regulations requires that water intended for human consumption should be taken from the highest quality source that is economically feasible. Accordingly, in assessing the adequacy of water being considered for potable reuse, comparison should be made with the highest quality water that can be obtained from that locality, even though that source may not be in use. Thus, from a microbiological point of view, the risk of infectious disease being transmitted by conventionally treated water has been minimized. But using a water source of inferior quality, such as wastewater, makes it necessary to reexamine the current microbiological and chemical criteria and standards and, where possible, to suggest alternatives.

This report addresses in detail some general characteristics of wastewater reuse systems, the chemical and microbiological constituents of such systems, concentration methods for analysis and toxicity testing, a methodology for health effects testing of reused water, and strategies for assessing and monitoring water quality for human exposure. It concludes with an assessment and criteria for potable water reuse.

REFERENCES

National Academy of Sciences. 1977. Drinking Water and Health. Safe Drinking Water Committee, Advisory Center on Toxicology, Assembly of Life Sciences, National Research Council. National Academy of Sciences, Washington, D.C. 939 pp.

National Academy of Sciences. 1980. Drinking Water and Health, Vol. 3. Safe Drinking Water Committee, Board on Toxicology and Environmental Health Hazards, Assembly of Life Sciences, National Research Council. National Academy Press, Washington, D.C. 415 pp.

2
Wastewater Reuse Systems

A nearly unlimited number of system alternatives are available for reclaiming wastewaters, and they cannot all be discussed here. This report is limited to the general efficiency and reliability of advanced treatment for contaminant removal, even though the effectiveness of various segregation and blending schemes is best considered in the context of a given situation.

ADVANCED WASTEWATER TREATMENT SYSTEMS

Considerable research has been conducted in both laboratories and in pilot plants to assess the various processes commonly considered for inclusion in advanced wastewater treatment plants reclaiming water for potable reuse. However, experience with full-scale operation of the processes is limited, especially when the processes are used in combination in a complete system.

Many process modifications and overall system alternatives for wastewater reclamation exist, but economics and process effectiveness have generally reduced the treatment alternatives to just a few. One treatment alternative, representative of many systems where reuse for potable purposes is contemplated or where high levels of treatment are required for other uses, is that used at Water Factory 21, a 57,000-m^3/day (15-million gal/day) advanced wastewater treatment system in Fountain Valley, Calif., operated by the Orange County Water District (Cline, 1979; McCarty et al., 1980). The processes included in this system are shown in Figure 2-1.

Water Factory 21, which became operational in 1976, was designed to provide water for injection into a heavily used groundwater supply to prevent the intrusion of seawater in the freshwater system. Local authorities required an advanced level of treatment. It was also necessary to prevent clogging of the aquifer near the injection points (Cline, 1979). At Water Factory 21, a portion of the injected water is likely to flow inland and become mixed with groundwaters being used elsewhere, thus providing some measure of indirect reuse.

Normal biological wastewater treatment, as required for municipal wastewater treatment in the United States, serves as pretreatment to most advanced treatment plants. At Water Factory 21, the water is pretreated by the Orange County Sanitary District (OCSD) and includes primary settling to remove a large portion of the suspended material

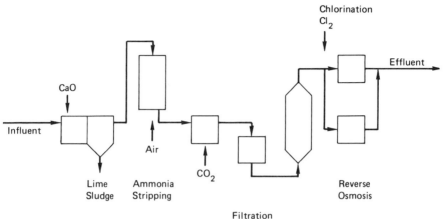

FIGURE 2-1 Processes and sampling locations at Water Factory 21 in Fountain Valley, Calif. From McCarty et al., 1978.

and secondary biological treatment with processed activated sludge to remove additional suspended solids and also soluble, biodegradable organic materials.

Advanced wastewater treatment generally includes additional removal of suspended material by chemical coagulation with lime, alum, or a ferric salt. This process is generally quite effective in removing heavy metals as well as dissolved organic materials. Air stripping is used at Water Factory 21 to remove ammonia, but it has also been found to be very effective for removing volatile organic compounds such as trihalomethanes, trichloroethylene, tetrachloroethylene, chlorinated benzenes, and lower molecular weight hydrocarbons (McCarty et al., 1980), for which other processess have been ineffective. Water Factory 21 uses lime; recarbonation by the addition of carbon dioxide is used to neutralize the resulting high pH. This step is followed by disinfection with chlorine and filtration to remove additional suspended solids that might clog the granular activated carbon (GAC) beds, which serve to remove additional soluble organic materials.

At Water Factory 21, a portion of the water stream then passes through reverse osmosis, which is used for demineralization so that (when blended back with the remaining water) the mixture will meet total dissolved solids requirements specified for injected water. The reverse osmosis process is also effective for removing approximately 90% of the remaining organic material. The blended water then receives chlorination for final disinfection prior to injection into the groundwater.

Table 2-1 lists several other advanced treatment systems with capacities of 3,300 m^3/day (1 million gal/day) or greater, together with the purpose of the facility and the processes used for treatment. The processes are generally similar to those used at Water Factory 21.

One of the earliest planned wastewater reuse schemes to help augment a potable water supply was that used at Whittier Narrows, Calif. When the plant was first operated, the water was treated by activated sludge and placed in spreading basins to percolate through the ground and become part of the groundwater aquifer used as a source for the local water supply. Surface water from other sources was percolated in the same basins so that the reclaimed water was diluted. Recently, the plant was modified to include filtration. Other similar plants providing indirect potable reuse have been built or are planned in the Los Angeles and Orange County areas (Horne, 1979; Nellor et al., 1979; Wassermann and Radimsky, 1979). Reclaimed water in these two areas has amounted to as much as 23% of the water entering the groundwater basin (Nellor et al., 1979).

In 1968 the first large-scale plant using current advanced treatment techniques was put into operation at South Lake Tahoe. The product water was not intended for potable reuse; it became a recreational lake used for fishing and boating. Nonetheless, the operation has provided experience with potable reuse technology. The processes used are similar to those at Water Factory 21, except there is no demineralization.

In 1969, the first (and, in 1981, the only) facility to provide direct potable reuse of wastewater began operating at Windhoek, Namibia. Modifications to improve influent quality and to reflect better technology were completed in 1976 (van Vuuren et al., 1980). The reclaimed water is pumped directly into the water treatment plant, blended with stored surface water, and treated by normal processes of coagulation and filtration. The reclamation plant was built to offset a serious water deficiency in the area, and the water provided at times comprises 20% to 50% of the municipal supply.

Pretreatment consists of trickling-filter biological treatment and storage in "maturation ponds," similar to the oxidation ponds with algae sometimes used in the United States. Before 1976, the algae in the maturation ponds were depended on to remove ammonia and were themselves removed in advanced treatment by air flotation. In 1976, lime treatment and air stripping were incorporated to replace the flotation units. The process stream is now similar to that at Water Factory 21.

The Stander experimental waste reclamation plant in Daspoort, Pretoria, South Africa, was designed as a full-scale plant to evaluate different processes and to provide information on efficiency and reliability of advanced treatment for potable reuse (Prinsloo et al., 1978). Water is a scarce resource in this area, and situations similar to that at Windhoek were anticipated. Between 1970 and 1976, the influent was trickling-filter treated municipal wastewater; since then, the influent has been treated by activated sludge. Similar changes were used later at the Windhoek plant.

The reclamation plant in Palo Alto, Calif. (operated by the Santa Clara Valley Water District) was also constructed to provide a

TABLE 2-1 Advanced Treatment Plants for Wastewater Reuse

Location	Date Started	Capacity (m³/day)	Purpose	Pre-treatment[a]	Advanced Treatment[b]	References
Whittier Narrows, Calif.	1962	47,000	Groundwater, recharge basins	AS	Filt, Cl_2	Horne, 1979; Parkhurst and Garrison, 1963
South Lake Tahoe, Calif.	1968	28,000	Recreational lake	AS	Lime, Air, CO_2, Filt, GAC, Cl_2	Culp et al., 1979
Windhoek, Namibia (formerly South-West Africa)	1969 (mod. 1978)	4,800	Direct potable reuse	TF, MP	Lime, Air, CO_2, Cl_2, Filt, Cl_2, GAC, Cl_2	van Vuuren et al., 1980
Daspoort, South Africa	1970	4,500	Experimental	AS	Lime, Air, CO_2, Filt, Cl_2, GAC, Cl_2	Prinsloo et al., 1978
Orange County, Calif.	1976	57,000	Groundwater injection	AS	Lime, Air, CO_2, Cl_2, Filt, GAC, RO, Cl_2	Cline, 1979; McCarty et al., 1980
Palo Alto, Calif.	1977	7,600	Experimental groundwater injection	AS	Lime, Air, CO_2, O_3, Filt, GAC, Filt, Cl_2	Fowler, 1979; Roberts et al., 1979

Location	Year	Capacity	Application	Treatment[a]	Processes[b]	Reference
Tahoe-Truckee, Calif.	1978	18,300	Water supply stream discharge	AS	Lime, CO_2, Filt, GAC, Ion, Land	Smith, 1979
Fairfax County, Va.	1978	57,000	Water supply reservoir discharge	AS	Lime, CO_2, Filt, GAC, Ion, Cl_2	Robbins and Gunn, 1979
Washington, D.C.	1981	3,800	Experimental direct potable reuse	AS	Air, Alum, Cl_2, Filt, GAC, Cl_2	Johnson and Aukamp, 1979
Denver, Colo.	1982 (est.)	3,300	Experimental direct potable reuse	AS	Lime, CO_2, Filt, Ion, Cl_2, GAC, O_3, RO, ClO_2	Rothberg et al., 1979; Work et al., 1980

[a] AS = activated sludge
TF = trickling filter
MP = maturation ponds
[b] Lime = lime treatment
Air = air stripping
CO_2 = recarbonation
Cl_2 = chlorination
O_3 = ozonation
ClO_2 = chlorine dioxide disinfection
Filt = filtration
GAC = granular activated carbon
RO = reverse osmosis
Alum = alum coagulation and settling
Ion = ion exchange
Land = land application

barrier to seawater intrusion. Demineralization was not required at Palo Alto because the water was injected into an already salty aquifer, but ozonation was provided to evaluate the usefulness and costs of an alternative disinfectant and to assess the possible advantages for removal of organic compounds, as reported by the Safe Drinking Water Committee (National Academy of Sciences, 1980). Filtration was provided both before and after GAC treatment to prevent clogging of the GAC system and to remove suspended biological particles produced within that system. At Palo Alto, the injected water does not blend with a usable water supply.

Two other advanced wastewater systems of similar design were started in 1978--one near Lake Tahoe (at the City of Truckee, Calif.) and the other in Fairfax County, Va. Both systems discharge into surface waters used downstream for local water supplies; thus, the reclaimed water is destined for some measure of indirect reuse. The two systems include ion exchange, using clinoptilolite for selective ammonia removal.

The Tahoe-Truckee effluent is percolated through the soil before it is allowed to enter the Truckee River, which is used as a water supply by Reno, Nev., 56 km downstream. The stringent treatment requirements were designed to help maintain the Truckee River as a pristine river, as well as to protect water users in Reno (Smith, 1979).

The treatment plant in Fairfax County is designed to protect water quality in the Occoquan Reservoir, which is the principal raw water source for the Fairfax County Water Authority, serving more than 660,000 people (Robbins and Gunn, 1979). During extended dry periods, the wastewater discharge constitutes the majority of flow into the resevoir. In the drought of September 1977, 80% of the flow to the reservoir came from this source; as the population increases, the proportion may become even greater. Thus, the quality of the reclaimed water destined for indirect reuse is of concern.

Operation of the Blue Plains experimental estuary water treatment plant in Washington, D.C., began in March 1978. The main purpose of this facility is to evaluate the feasibility of using the Potomac River estuary as a possible water source for the metropolitan area. Construction and operation of the plant were authorized by Congress in 1974 (Johnson and Aukamp, 1979). However, there is concern regarding the quality of estuary water during future droughts--when the estuary may contain up to 50% effluent from treatment plants discharging filtered secondary treatment municipal wastewater into the estuary as well as runoff water from a highly urbanized area. Because of the relative proximity of the wastewater discharge points and possible water intakes, it is not clear whether this situation should be considered as direct or indirect reuse.

The processes to be evaluated in this facility are similar to the treatment systems already discussed, although aeration at Blue Plains is minimal and may not be as effective in removing volatile organic compounds as the aeration provided at Water Factory 21. In addition to the processes listed in Table 2-1, smaller scale studies are planned for the demineralization of sidestreams by reverse osmosis and also by electrodialysis. Future quality of the Potomac estuary (during droughts) is being simulated by blending estuary water and filtered secondary effluent from the Blue Plains Municipal Treatment Plant.

Another experimental treatment plant is scheduled to begin operation in 1982 in Denver, Colo. (Rothberg et al., 1979; Work et al., 1980). This plant is designed to evaluate the feasibility of direct potable reuse of treated municipal wastewater to help offset an anticipated water shortage. The plant will use ion exchange to remove ammonia and will treat sidestreams by reverse osmosis. Chlorine dioxide is now planned for disinfection to reduce the quantity of chlorinated organic materials formed when chlorine is used. The extensive toxicological testing program at Denver should add significantly to the knowledge of health risks associated with wastewater reuse for potable purposes.

ADVANCED WASTEWATER TREATMENT COSTS

The series of processes used at the advanced wastewater treatment plants can remove contaminants quite effectively, but at substantial costs. To provide some perspective, the experience from Water Factory 21 is summarized here and compared with the costs of municipal water supplies in general. The costs of treatment at Water Factory 21 should not, however, be considered directly applicable to those of other locations because of differences in many local factors, including process design and cost of construction, labor, land, and material.

Table 2-2 summarizes the construction and operating costs for Water Factory 21 for 1 year (Argo, 1980). Construction costs were amortized over a 20-year period at 7% interest. Reverse osmosis is highly effective in removing both inorganic and organic contaminants but, as shown in the table, costs more to operate than do all the other processes combined.

As a comparison, the average 1974 treatment cost for 12 major U.S. cities was $9.2 per 1,000 m^3, and the average total cost, including support services, pumping, and distribution, was $110 per 1,000 m^3 (Clark, 1979). These figures illustrate that reclaimed wastewater is expensive and likely to be economically feasible only where water is scarce and not available at what might be considered a normal cost.

Morever, the operating staff and monitoring personnel need to have a higher level of training than is now generally found in facilities of this type. Furthermore, considerably more effort must be made in preparing personnel for this work if it is ever to become part of the public water supply. Needless to say, this will also add to the overall cost.

EFFECTIVENESS OF STORAGE

Additional removal of contaminants can occur during storage of reclaimed wastewaters in reservoirs or aquifers. Many of the identifiable contaminants in wastewaters are volatile and lost to the air in open reservoirs. Others can be photochemically transformed by

TABLE 2-2 Cost of Advanced Wastewater Treatment at Water Factory 21 for 1 Year[a]

Type of Treatment	Cost (Dollars/1,000 m^3)		
	Capital	Operations and Maintenance	Total
Lime treatment and recarbonation	$17.5	$35.7	$53.2
Air stripping[b]	15.6	15.2	30.8
Mixed-media filtration	4.8	2.9	7.6
GAC treatment and regeneration	16.2	19.7	35.9
Chlorination	1.2	11.1	12.2
SUBTOTAL	$55.3	$84.6	$139.7
Reverse osmosis	45.3	118.5	163.8
Injection	4.5	9.2	13.7
TOTAL[c]	$105.1	$212.3	$317.2

[a]From Argo, 1980.
[b]With fans operating.
[c]Assuming all water treated by reverse osmosis.

exposure to sunlight. Pathogens generally decrease in number during storage as a result of natural death and predation. All of these positive effects result from storage of reclaimed waters in open reservoirs. Unfortunately, a negative effect is the growth of algae, which is likely to occur because remaining inorganic nutrients produce soluble organic exudates. These exudates affect the color, taste, and odor of the water and also react with chlorine to produce chlorinated organic compounds. Such negative features of storage in open reservoirs may offset the water quality gains and must be considered during the design phase.

Ground storage can circumvent many of these problems, but it does not provide all of the advantage of surface storage. Because groundwaters are not in contact with air, losses by volatilization and photolysis are unlikely. The major benefits from passage of reclaimed waters through the ground, by either percolation or injection, are the adsorption, ion exchange, and opportunity for biological transformation afforded. Adsorption will not provide a long-term solution, but it can remove many hydrophobic contaminants quite effectively from reclaimed water for several years before breakthrough occurs. Ground storage can also effectively reduce concentrations of bacteria, viruses, and heavy metals. Aquifer storage of reclaimed waters tends to improve their quality, and this method should be considered. The potential blending with native groundwaters to reduce contaminant concentrations can also be an advantage. However, many (and perhaps most) of the remaining contaminants in reclaimed wastewaters are not effectively removed by passage through the ground, and many of the organic materials formed from wastewater chlorination are not sufficiently hydrophobic for their movement through the ground to be impeded. In addition, the major portion of the effluent of organic materials from advanced wastewater treatment systems are those that have effectively escaped removal by biological, physical, and chemical processes and, thus, are likely to by unaffected by similar processes operating in the ground. The nature and health significance of these remaining materials must be considered. Once contaminated, groundwater is difficult to cleanse. Thus, added precautions must be taken when introducing water containing any contaminants.

REFERENCES

Argo, D.G. 1980. Cost of water reclamation by advanced wastewater treatment. J. Water Pollut. Control Fed. 52:750.

Clark, R.M. 1979. Labor wage rates, productivity, and the cost of water supply. J. Am. Water Works Assoc. 71:364.

Cline, N.M. 1979. Groundwater recharge at Water Factory 21. Proc. Water Reuse Symp. 1:139. American Water Works Association Research Foundation, Denver, Colo.

Culp, G., D. Hinrichs, R. Williams, and R. Culp. 1979. Wastewater reuse alternatives at South Lake Tahoe. Proc. Water Reuse Symp. 1:595. American Water Works Association Research Foundation, Denver, Colo.

Fowler, L.C. 1979. Santa Clara Valley Water District's reclamation facility at Palo Alto. Proc. Water Reuse Symp. 1:179. American Water Works Association Research Foundation, Denver, Colo.

Horne, F.W. 1979. Regional water reuse in Southern California. Proc. Water Reuse Symp. 1:546. American Water Works Association Research Foundation, Denver, Colo.

Johnson, C.C., and D.R. Aukamp. 1979. The experimental estuary water treatment plant. Proc. Water Reuse Symp. 1:613. American Water Works Association Research Foundation, Denver, Colo.

McCarty, P.L., M. Reinhard, C. Dolce, H. Nguyen, and D.G. Argo. 1978. Water Factory 21: Reclaimed Water, Volatile Organic, Virus, and Treatment Performance. EPA 600/2-78-076. Municipal Environmental Research Laboratory, Office of Research and Development, U.S. Environmental Protection Agency, Cincinnati, Ohio. 87 pp.

McCarty, P.L., M. Reinhard, J. Graydon, J. Schreiner, K. Sutherland, and T. Everhart. 1980. Advanced treatment for wastewater reclamation at Water Factory 21. Municipal Environmental Research Laboratory, U. S. Environmental Proctection Agency, Cincinnati, Ohio.

National Academy of Sciences. 1980. Drinking Water and Health, Vol. 2. Safe Drinking Water Committee, Board on Toxicology and Environmental Health Hazards, Assembly of Life Sciences, National Research Council. National Academy of Sciences, Washington, D.C. 393 pp.

Nellor, M.H., F.D. Dryden, and Ching-lin Chen. 1979. Health effects of groundwater recharge. Proc. Water Reuse Symp. 3:2146. American Water Works Association Research Foundation, Denver, Colo.

Parkhurst, J.D., and W.E. Garrison. 1963. Water reclamation at Whittier Narrows. J. Water Pollut. Control Fed. 35:1094.

Prinsloo, J., S.H.V. van Blerk, and J. van Leeuwen. 1978. Comparative reclamation of potable water from biofilter and activated sludge effluents at the Stander Water Reclamation Plant. Prog. Water Technol. 10(1-2):81.

Robbins, M.H., Jr., and G.A. Gunn. 1979. Water reclamation for reuse in Northern Virginia. Proc. Water Reuse Symp. 2:1311. American Water Works Association Research Foundation, Denver, Colo.

Roberts, P., P.L. McCarty, and W.M. Roman. 1979. Direct injection of reclaimed water into an aquifer. J. Environ. Eng. Div. Am. Soc. Civil Eng., EE4, 105:675.

Rothberg, M.R., S.W.W. Work, K.D. Linstedt, and E.R. Benjnett. 1979. Demonstration of potable water reuse technology, the Denver Project. Proc. Water Reuse Symp. 1:105. American Water Works Association Research Foundation, Denver, Colo.

Smith, S.A. 1979. Tahoe-Truckee water reclamation plant first year in review. Proc. Water Reuse Symp. 2:1435. American Water Works Association Research Foundation, Denver, Colo.

van Vuuren, L.R.J., A.J. Clayton, and D.C. van der Post. 1980. Current status of water reclamation at Windhoek. J. Water Pollut. Control Fed. 52:661.

Wassermann, D., and J. Radimsky. 1979. Water reclamation efforts in California. Proc. Water Reuse Symp. 1:69. American Water Works Association Research Foundation, Denver, Colo.

Work, S.W., R.R. Rothberg, and K.J. Miller. 1980. Denver's potable reuse project: Pathway to public acceptance. J. Am. Water Works Assoc. 72:435.

3

Chemical and Microbiological Constituents of Reuse Systems

CHEMICALS

In attempting to develop criteria and standards for chemical constituents of treated wastewater intended for potable reuse, it is not sufficient to consider only the "simple" question of whether specific chemicals will or will not exceed health-related maximum contaminant levels (MCL's). On the contrary, a broad data base is needed in order to determine whether the adverse effects of treated wastewater on human health are greater than those presented by "natural" (and even anthropogenically contaminated) waters used as sources for our domestic supplies. This data base might include information on the nature of chemicals in the (domestic) wastewater to be treated and measurements of their concentrations over time; the efficacy of the treatment of these constituents; and, ultimately, the concentrations of these chemicals that will be present in the treated wastewater eventually delivered to the consumer for potable use. The treatment steps themselves are important in that they, in effect, replace what would otherwise be natural processes, e.g., dilution, precipitation, and sorption.

With respect to the municipal wastewater being treated, it is useful to characterize the possible sources of chemicals in the first water use (domestic and possible industrial and runoff inputs), the impact of conventional (usually secondary) wastewater treatment, and the typical concentrations of the inorganic chemicals and their variability in the secondary effluent to be used as a source. The possible buildup of "macro" inorganics (such as sulfate, calcium, and magnesium), although they may not have adverse health implications, may be esthetically undesirable. In addition, many trace inorganics (such as nitrate, mercury, and lead) can affect health adversely. Analyses of the sources of various constituents and possible increases in their concentration may show that it is necessary to limit industrial inputs strictly to the wastewater system or perhaps to reuse only wastewater that receives very few or no substantial industrial effluent.

Several questions arise regarding the presence and possible buildup of chemical constituents in wastewater to be used for a potable water supply.

- Which chemicals are likely to occur in treated wastewater at concentrations substantially higher than those currently found in acceptable public water supplies?
- Is existing information on the composition and effects of constituents in treated domestic wastewater and actual or pilot plant wastewater treatment systems sufficient to provide criteria on potability of reused wastewater?
- Are domestic wastewater and wastewater treatment systems sufficiently alike to permit a confident assessment of reuse, based on studies of only a few systems?
- What is the extent of the variabilities in the quality of treated municipal effluents over time, as well as in the outputs of the treatment systems?

Answers to these questions provide a useful and necessary perspective to assessment of any unusual or high exposures to chemicals from the ingestion of treated wastewaters and their possible adverse effects on human health.

Inorganic Substances

Public water supplies accumulate a variety of inorganic substances as a result of different domestic uses. Table 3-1 shows the typical ranges of added concentrations for some minerals, but not for the wide variety of trace inorganic chemicals (typically those at concentration of <1 mg/liter) that may also enter public water supplies

TABLE 3-1 Some Minerals That Enter Public Supplies During Domestic Use[a]

Constituent	Ranges of Increases (mg/liter)
Boron	0.1-0.4
Sodium	40-70
Potassium	7-15
Magnesium	4-10
Calcium	6-16
Phosphate (PO_4^{-3})	20-40
Sulfate (SO_4^{-2})	15-30
Chloride (Cl^-)	20-50
Total dissolved solids	100-300
Alkalinity (as $CaCO_3$)	100-150
Total nitrogen (nitrate, ammonia, organic)	20-40

[a]From Chang and Page, 1980.

after domestic and industrial uses. Thus, in analyzing municipal wastewater for possible reuse, a comprehensive study should include not only an assessment of possible health effects from reusing domestic wastewater, but also a measurement of the amount and quality of industrial waste in the system and (if the system receives storm water) runoff constituents.

A number of potentially toxic elements may be present in municipal wastewater and sewage; these include arsenic, cadmium, chromium, copper, lead, mercury, selenium, and silver. Evaluation of the potential health consequences from such elements is complicated by the possibility that each might exist in a number of different chemical states, differing in solubility, reactivity, and toxicity. Thus, a metal cation may have more than one oxidation state, form a number of complexes and chelates with organic ligands, or form organic metallic compounds. Furthermore, interaction between two or more metal cations may modify toxicity in one direction or another.

All the processes involved in wastewater treatment can affect the bioavailability and toxicity of metals. Many metal cations such as those of cadmium (Cd^{+2}), copper (Cu^{+2}), lead (Pb^{+2}), mercury (Hg^{+2}), and monomethyl mercury (CH_3Hg^+) form tight bonds with organic ligands present in most particulate matter. Microorganisms tend to assimilate metals from water, which can lead to their removal. The production of hydrogen sulfide by certain microorganisms can cause precipitations of cadmium sulfide, mercuric sulfide, and lead sulfide; in addition, some microorganisms can change the oxidation state of the metal. Thus, Hg^{+2} can be reduced to the metallic Hg^0, which can volatilize from the system.

A few potentially toxic metals, e.g., arsenic, lead, mercury, and tin, can form stable organometallic compounds, which can be produced anthropogenically (mainly from industrial sources) or by microorganisms. Microorganisms can both synthesize and break down organic metallic compounds. The best are the organic mercury compounds, which are used as fungicides, released into the environment, and are broken down to inorganic Hg^{+2}. When this cation is present in sediment in water, it becomes the substrate for certain methanogenic bacteria and is converted to monomethyl (CH_3Hg^+) or dimethyl (CH_3HgCH_3) mercury. Methyl mercury has been shown to be toxic to the human nervous system (National Academy of Sciences, 1977).

The outcome of bacterial synthesis and degradation of organic metallic compounds will be affected by the type of metal, the type of bacteria, and ambient conditions such as oxygen tension and pH. Although methyl mercury is highly toxic to humans, it is unlikely to attain hazardous concentrations in potable water because of its low solubility in water and because it adsorbs on particulates. Dimethyl mercury, preferentially produced under alkaline conditions, should vaporize into the atmosphere.

A number of studies have shown that secondary or biologically treated municipal wastewater can contain a wide range of trace element concentrations, as shown in Table 3-2. For many of the constituents, the range concentrations may differ by as much as a factor of 100, presumably reflecting the variation both among and within secondary effluents. It is thus unlikely that one can define a "typical" municipal secondary effluent with respect to its

TABLE 3-2 Concentrations of Trace Elements Found in Treated Wastewater and Water Quality Criteria for Public Water Supplies and Irrigation Water[a]

Element	Wastewater Effluent (mg/liter) Range	Median	Water Quality Criteria (mg/liter)		
			Public Water Supplies	Irrigation Water	
				Continuous Use[b]	Short-Term Use[c]
Arsenic	0.005-0.023	0.005	0.1	0.1	2.0
Boron	0.3 -2.5	0.7	--	0.75	2.0
Cadmium	0.005-0.22	0.005	0.01	0.01	0.05
Chromium	0.001-0.1	0.001	0.05	0.1	1.0
Copper	0.006-0.053	0.018	1.0	0.2	5.0
Lead	0.003-0.35	0.008	0.05	5.0	10.0
Molybdenum	0.001-0.018	0.007	--	0.01	0.05
Mercury	0.0002-0.001	0.0002	0.002	--	--
Nickel	0.003-0.60	0.004	--	0.20	2.0
Selenium	--	--	--	0.2	0.02
Zinc	0.004-0.35	0.04	0.05	2.0	10.0

[a]From Chang and Page, 1980.
[b]For water used continuously on all soils.
[c]For use up to 20 years on fine-textured soils of pH 6.0 to 8.5.

inorganic constituents, nor can one predict their concentrations in a specific system being considered for reuse. In this regard, the variability and mean concentrations of inorganic chemicals in a secondary effluent are also of concern.

Table 3-3 presents such data for the secondary effluent used as the source water at Water Factory 21 (McCarty et al., 1980). The constituents listed are those specified in the National Interim Primary Drinking Water Regulations (Environmental Protection Agency, 1976). The geometric mean concentrations for many of the inorganic constituents vary substantially between the two periods studied, during which different sources of wastewater were used. For example,

TABLE 3-3 Comparison Between the Maximum Contaminants Levels (MCL's) Recommended in the National Interim Primary Drinking Water (NIPDW) Regulations and Levels Found in Influent Water at Factory 21[a]

Contaminant	MCL	Influent Water			
		First Period Studied		Second Period Studied	
		Geometric Mean	98% of Time Less Than:	Geometric Mean	98% of Time Less Than:
mg/liter:					
Arsenic	0.05	0.005	0.005[b]	0.005	0.005[b]
Barium	1.0	0.08	0.14	0.03	0.06
Cadmium	0.01	0.026[c]	0.07	0.033	0.15
Chromium	0.05	0.14	0.31	0.048	0.11
Lead	0.05	0.02	0.051	0.007	0.017
Mercury	0.002	0.0016	0.025	0.001	NK[g]
Nitrate (as N)	10	0.23	1.2	2.8	49
Selenium	0.01	0.0025	0.0025[b]	0.0025	NK[g]
Silver	0.05	0.003	0.007	0.001	0.006
Fluoride	1.4[d]	1.4	2.0	1.3	1.9
μg/liter:					
Endrin	0.2	0.01	0.01	0.01	0.01
Lindane	4	0.2	0.9	0.14	0.22
Toxaphene	5	0.01	0.01	0.01	0.01
2,4-D	100	0.01	0.01	0.01	0.01
2,4,5-TP	10	0.01	0.01	0.01	0.01
Methoxychlor	100	0.1	0.1	0.01	0.01
MPN[e]/100 ml:					
Coliforms	1	89	38,000	1.6	195
TU[f]:					
Turbidity	1	42	79	7	54

[a]From McCarty et al., 1980.
[b]Based on less than 1 in 20 samples analyzed.
[c]Underlined values represent those exceeding MCL's.
[d]Temperature = 26.3°C.
[e]MPN = Most probable number.
[f]TU = Turbidity units.
[g]Not known.

the two mean values for lead are 0.02 and 0.007 mg/liter. Another important measure in the table is the highest concentration—which is not exceeded 98% of the time. For some of the constituents, this concentration is much higher than that of the geometric mean during the same time period. For example, the concentrations for mercury during the first period were 0.025 and 0.0016 µg/liter, respectively—a factor of approximately 15. From such data, it is clear that the variability of a secondary sewage effluent over time may affect the quality of the renovated water product.

The removal efficiencies of advanced wastewater treatment processes can be highly variable, with respect to both the process and the inorganic constituents it removes. Englande and Reimers (1979) have reviewed the technical literature and the results from specific advanced wastewater treatment systems. Their findings for several trace elements and other constituents are summarized in Table 3-4. Not only the specific processes affect various elements differently, but also other water quality variables and influent concentrations play roles as well. Clearly, no single process (at least among those summarized) readily stands out as the clear choice for "general" trace element removal. Even reverse osmosis (RO), which would be expected to be highly efficient in removing inorganic constituents, was found to be much less effective for most trace elements than for "macro" inorganics (Hrubec et al., 1979).

When Water Factory 21 began operation in October 1976, the influent to this advanced wastewater treatment (AWT) plant was trickling-filter treated municipal wastewater from the Orange County Sanitary District. In March 1978, the influent water was upgraded to activated sludge-treated municipal wastewater, which was then segregated to reduce the contamination with industrial wastes prior to AWT. This procedure reduced concentrations of aromatic hydrocarbons, phthalates, and chemical oxygen demand (COD).

Tables 3-5 and 3-6 compare the influent and effluent concentrations of some heavy metals and other contaminants generally present in sufficient quantities to be measured with reasonable precision. The AWT effluent concentrations are those measured after treatment by all removals processes except RO. The values for AWT plus RO represent the effect of all processes except final chlorination.

The constituents listed in these tables are commonly found in wastewater and contaminated surface water supplies, although the percentage removals obtained differ for each contaminant. The tables also list the particular process or processes that were most effective in obtaining this level of removal. Chemical precipitation with lime (ppt), granular activated carbon (GAC), and RO treatment were particularly effective for removing the collective organic parameters of COD and total organic carbon (Table 3-5). In general, heavy metals were removed most effectively by chemical precipitation, although several were also removed by GAC and RO treatment. Air stripping was effective in ammonia removal and also in the removal of many of the chlorinated compounds; in fact, stripping tended to be more effective for most individual compounds. The results indicate that a combination of processes is required to achieve good overall removal of the many different contaminants in municipal wastewaters.

The variability for trace inorganics over time has also been considered in some detail in the Water Factory 21 study by McCarty et al. (1980), in terms of both variability in influent concentration and removal efficiency. The results of that study (for eight heavy metals) are shown in Figure 3-1. The results are plotted as log distributions, which are often encountered and which were found to be a useful way to consider the variabilities over time. However, confidence limits or uncertainties are associated with the log normal distribution lines. Parallel distributions for the influent and effluent concentrations are consistent with the fractional removal and are independent of concentration; however, the lines for the metals typically are not parallel.

The results for these elements and a few others are shown in Table 3-7, which presents not only the average percent removals but also the 95% confidence interval for the average. The interval can be quite great for some elements, such as copper, lead, and silver. This scale offers a conservative way to predict average maximum concentrations in the effluent, based on the minimum average percentage of removal. Thus for cadmium, 47% average removal should be used rather than the 84% (the two extremes of the 95% confidence band).

Clearly many variabilities are encountered in effluents from secondary and advanced waste treatment processes, and it is difficult to predict removals a priori for a given treatment system. Thus, as in the Water Factory 21 study, it may be necessary to evaluate the full range of variabilities over an extended period.

Radionuclides

As a result of discharges, radionuclides can appear in increased concentrations in treated wastewater, as compared to their presence in potable water at first use. Regulated events include discharges of the radiochemicals used in research and then discharged into municipal water systems; unregulated discharges are typified by the substantial and largely unknown amounts of radioisotopes given to patients for medical diagnostic and therapeutic purposes and then discharged into sewers. At one large sewage treatment plant, such medical radioactive wastes were shown to contribute concentrations of iodine-131 and technetium-99m amounting to a total activity of 13 to 15 pCi/liter in the treated effluent (Moss, 1973).

Wastewater treatment processes can be expected to remove radionuclides at least partially (Environmental Protection Agency, 1977). This study noted that RO removed more than 99% of the radioactivity from a low-level radioactive laundry waste. Because many of the manmade and natural radionuclides in sewage are likely to exist in a simple inorganic ion form in domestic sewage, some of the wastewater processes that remove inorganic ions should be effective for them as well, including RO and ion exchange. Nevertheless, because concentrations could build up to potentially harmful levels in treated wastewaters, radioactivity should be monitored in such systems.

Monitoring requirements and maximum contaminant levels (MCL's) are specified in the U.S. National Interim Primary Drinking Water

TABLE 3-4 Pollutant Removal by Wastewater Treatment Processes[a]

Constitutent	Secondary Treatment (Biological)	Chemical Precipitation			Activated Carbon Adsorption	Comments on Activated Carbon	Residual Concentration Level (µg/liter)	General Comments on Removal
		Lime	Ferric Chloride	Alum				
Total dissolved solids	P	P	--	--	P	--	--	Generally increased total dissolved solids with treatment; reverse osmosis effective in removal
Ammonia nitrogen	VG	P	--	--	P	--	--	Biological nitrification most effective; breakpoint chlorination and stripping towers F to VG
Nitrate nitrogen	VG	P	--	--	P to G	Depends on anaerobic bioactivity	--	Biological denitrification most feasible
Phenol	P	P	--	--	P	Limited by driving force to about 1 mg/liter	>1	Treatment methods effective in reducing phenol to 1 µg/liter limit
Trace organics	G	G	--	--	G	Removal depends on specific organics	5,000 (total organic carbon and chemical oxygen demand)	Chlorinated organics may be increased with breakpoint chlorination; ammonia stripping effective in removing volatile refractory organics
Arsenic	P to F	P to G	G	--	P	Reacts with sulfide	3	Depends on influent level, pH, and redox potential
Barium	F	P to G	--	G	P to G	Due to highly soluble nature	>30	Enhanced precipitation as sulfate concentration increases

Metal								
Boron	P	P	--	--	P	--	>290	Generally negligible
Cadmium	P to G	F to VG	--	F	P to VG	Old carbon better	2	High removals due to precipitation of sulfide and hydroxide forms
Chromium	F to G	G	VG	$G(Cr^{+6})$ $VG(Cr^{+3})$	P to G (Cr^{+3}) $VG(Cr^{+6})$	Reduction with bioactivity Cr^{+3} less soluble than Cr^{+6}	20	Depends on influent level and oxidation state
Copper	F to G	P to G	--	G	G to VG	Enhanced sorption better with new carbon	70	Influenced by influent concentration
Iron	P to F	F to G	--	VG	P to G	Sulfide complexes ppt, but anaerobic bioactivity causes reduction to soluble Fe^{+2}	>40	Depends on influent level, pH, and redox potential
Lead	F to G	F to G	--	VG	P to G	--	>5	Enhanced precipitation with higher sulfate levels
Manganese	P	G to VG	--	P	P	Bioactivity on the carbon reduces Mn^{+4} to Mn^{+2}	5	Depends on pH and redox potential
Mercury	P to G	P to G	VG	G	P to G	Variability due to biological activity	5	Removal is a function of pH, initial concentration, and degree of complexation
Selenium	F	P to G	G	F	P to G	Variability due to highly soluble characteristics	2	Depends on influent concentration
Silver	P to G	P to VG	VG	VG	P to G	High affinity for sulfhydryl groups	2	Depends on influent level
Zinc	F to G	P to F	--	F	P to G	Zinc sulfide precipitate	>60	Depends on influent and sulfate levels

[a]From Englande and Reimers, 1979.

NOTE: P = Poor (<30%); F = Fair (30%–60%); G = Good (60%–90%); VG = Very Good (>90%).

TABLE 3-5 Selected Contaminant Concentrations and Effectiveness of Treatment at Water Factory 21 in Removing Them[a]

Contaminant or Other Factor Measured	Concentration as Noted			Percent Removal		Dominant Removal Processes[d]
	AWT Plant[b] Influent	AWT Plant[b] Effluent	RO[c] Effluent	AWT[b]	AWT[b] and RO	
Chemical oxygen demand (COD), mg/liter	47	12	1.3	74	97	ppt, GAC, RO
Total organic carbon (TOC), mg/liter	12	6	2.6	50	78	ppt, GAC, RO
Total dissolved solids (TDS), mg/liter	900	850	77	5	91	RO
Electroconductivity (EC), μS/cm	1,500	1,320	156	12	90	RO
Nitrogen, mg N/liter:						
Organic	2.0	1.1		45		ppt
Ammonia	4.0	0.8		80		strip, Cl$_2$
Nitrate	2.8	7.7	3.3	-175	-18	
Boron, mg/liter	0.74	0.53		28		
Fluoride, mg/liter	1.3	0.81		38		
Coliforms, MPN[e]/100 ml						
Total	1.6 × 10^6	0.05		100		ppt, Cl$_2$
Fecal	0.55 × 10^6	<1		100		ppt, Cl$_2$
Phenol, μg/liter	4.9	6.9		72		
Cyanide, μg/liter	25	0.8		98		
Color, units	37					
Methylene blue active substance (MBAS), mg/liter	0.25	0.08		68		

[a] From McCarty et al., 1980.
[b] AWT refers to all advanced wastewater treatment processes except reverse osmosis (RO).
[c] RO = Reverse osmosis.
[d] ppt = Chemical precipitation with lime.
   GAC = Granular activated carbon.
   Strip = Air stripping.
[e] MPN = Most probable number.

TABLE 3-6 Concentrations of Heavy Metals and Effectiveness of
Treatment at Water Factory 21 in Removing Them[a]

Contaminants	Concentration at AWT Plant (µg/liter)		Percent Removal AWT[b]	Dominant Removal Processes[c]
	Influent	Effluent		
Arsenic	<5	<5	--	
Barium	30	7.4	75	ppt, RO
Cadmium	33	9.5	71	ppt, RO
Chromium	48	3.1	94	ppt, GAC, RO
Copper	72	16	78	ppt, GAC, RO
Lead	7.1	1.0	86	ppt, GAC
Iron	98	42	57	ppt, RO
Mercury	<1	<1	--	
Manganese	29	1.7	94	ppt
Silver	1.2	0.7	42	ppt
Selenium	<5	<5	--	
Zinc	127	100	21	ppt, GAC

[a]From McCarty et al., 1980.
[b]AWT refers to advanced wastewater treatment processes except reverse osmosis (RO).
[c]ppt = chemical precipitation with lime.
GAC = granular activated carbon.

Regulations for radium-226, radium-228, gross alpha particle radioactivity, and beta particle and photon radioactivity from manmade radionuclides (Environmental Protection Agency, 1976). Thus, it is reasonable to require pilot-demonstration plants and potable wastewater treatment systems to have similar monitoring systems for radionuclides to ensure that the MCL's specified in the regulations are not exceeded. The regulations require at least quarterly monitoring of community water systems, which use waters contaminated by effluents from nuclear facilities. Depending on the quantities found, gross betas, strontium-89, strontium-90, cesium-134, iodine-131, and tritium must be monitored. Since there is a greater than normal likelihood that humans will be exposed to these isotopes, more frequent and more varied analyses are stipulated in the regulations.

Based on actual experience and on the potentially higher concentrations that can occur in reuse systems, there is a need for even more frequent monitoring than that specified, with a determination of specific isotopes (such as technetium-99m) likely to be found or to occur in high quantities. Unacceptably high quantities of radioactivity in otherwise acceptable treated water may be reduced to acceptable levels by the radioactive decay that occurs during a holding period before reuse. However, even more frequent monitoring

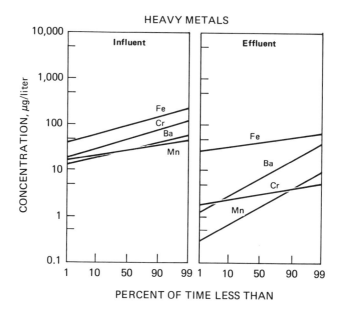
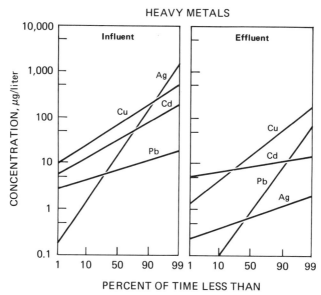

FIGURE 3-1 Distribution of heavy metal concentrations in the influent and effluent of Water Factory 21 from March 1978 to January 1979. From McCarty et al., 1980.

TABLE 3-7 Summary of Heavy Metal Concentrations and Removals by AWT During One Operating Period[a]

Heavy Metal	Geometric Mean Concentration (µg/liter)		Percent Removal	
	Influent	Effluent	Average	95% Confidence Interval for the Average
Arsenic	<5	<5	--	--
Barium	30	7.4	75	41 to 84
Cadmium	33	9.5	71	47 to 84
Chromium	48	3.1	94	90 to 96
Copper	72	16	78	16 to 94
Iron	98	42	57	40 to 70
Lead	7.1	1.0	86	-164 to 99
Manganese	2.9	1.7	94	86 to 97
Mercury	1	1	--	--
Selenium	5	5	--	--
Silver	1.2	0.7	42	-24 to 73
Zinc	127	<100	>21	--

[a]From McCarty et al., 1980.

of the system should then be required to ensure compliance with standards and the protection of human health.

Organics

Although quantitative information about the variability of major and trace inorganic constituents of wastewater to be treated, as well as the product, has often been well documented, similar data on organic compounds are scarce. However, there is a considerable amount of information about specific organic compounds detected in various waters with potential for potable reuse, but only limited data on such compounds in actual reclaimed waters. Most of the information on organic compounds has been reported only within the last decade, when important developments in analytical techniques and instrumentation provided data on specific compounds from water samples. In the early 1970's much qualitative (and very little quantitative) data were obtained.

There are lists of organic compounds present in various waters (e.g., municipal wastewaters, groundwater, surface waters, and drinking water), as well as the number of times the compounds were detected, and their concentrations, if known (Shackelford and Keith, 1976). Frequency-of-occurrence information often is inherently skewed because many studies have been designed to detect only a specific group of selected compounds. Even studies designed to detect and identify a broad range of organic compounds in a particular water sample produce incomplete frequency-of-occurrence data since analytical techniques used for most studies provide information only about compounds that can be tested by gas chromatography. Information about nonvolatile compounds is rarer.

Because of the large volume of current information, a comprehensive computerized data base to develop a list of organic compounds detected in various types of water is needed; the WATERDROP system (Garrison et al., 1979) is a positive step in this direction.

Of primary concern is whether identification has been confirmed or is only tentative. Quantitative data need to be evaluated with respect to recovery information and quantitation techniques. Qualitative information about occurrence of organic compounds known to affect health is useful because it may indicate a potential problem area.

Some of the best information about specific organic components in reclaimed water with potential for reuse was produced by the studies conducted at Water Factory 21 (McCarty et al., 1980). Other studies currently under way will provide additional information within the next 2 years (Battelle Columbus Laboratory, 1980; Environmental Protection Agency, 1980). Table 3-8 lists the organic compounds identified in the influent and effluent from Water Factory 21. More than 25 trace organic materials were identified in concentrations sufficient to allow quantification and estimates of removal efficiency. The average removal levels for some of these compounds are summarized in Table 3-9 for one time period. As shown, only the phthalates and some of the synthetic chlorinated organic compounds generally reached concentrations greater than 1 µg/liter, even in the influent. Air stripping was effective for most of these chemicals, except for the nonvolatile and hydrophobic organic materials such as lindane and polychlorinated biphenyls. The latter were removed to some degree by precipitation and granular activated carbon. Generally, removal of aromatic hydrocarbons was not great, but the results are uncertain because the concentrations were very low and and near detection limits. Before the changeover to activated sludge treatment, the aromatic compounds were found at higher concentrations and were then effectively removed by air stripping. As with removing inorganic compounds, a combination of processes is required to achieve effective removal of the aromatics.

In the various studies of wastewater treatment for potable reuse in South Africa, data have been reported on the presence of a number of trace organic chemicals in the finished water, and on the efficiency of removal. In a study conducted at the Stander Water Reclamation Plant, Stander (1980) observed that 10 different polynuclear aromatic hydrocarbons were detected in the ng/liter concentration range in the influent, and only 2 (pyrene and fluoranthene) in the final water. He noted that "the reclaimed water contains less organic substances and at lower concentrations than water purified by conventional treatment of raw surface waters receiving pollutants from non-point sources and from discharges of purified sewage and industrial effluent."

Van Rensburg et al. (1980) reported their results from monitoring

17 organic micropollutants injected into a scaled-down version of the Stander Water Reclamation Plant. The study was designed to determine the ability of the plant and its unit processes to remove a variety of toxic pollutants of possible industrial, agricultural, and domestic origin. Some of their results are shown in Table 3-10. For each of the compounds, the overall removal was greater than 99%. The authors concluded that the reclamation plants studied were "capable of effectively removing organic industrial and other pollutants in 'shock load' quantities."

MICROBIOLOGICAL CONSTITUENTS

Because of the potential for widespread waterborne disease, public officials have been concerned about biological contaminants in public water supplies for more than 100 years. Heretofore, it has been common practice to develop a public water supply using the highest quality raw water source available. For the most part, such sources have tended to minimize the risk of transmitting infectious diseases. Before using domestic wastewater, such as that in direct potable water reuse, or raw water significantly contaminated with wastewater for a public water supply, the potential for the spread of infectious diseases should be reexamined.

Acute microbial diseases are occasionally transmitted to broad segments of the population via the water supply. However, the assessment of or hazard associated with the transmission of infectious agents through water supplies is difficult. The presence of microorganisms in or on living things does not necessarily mean that the host will develop disease. Likewise, a waterborne disease usually occurs in a population without concomitant recovery of the organism from the water because of technical problems associated with the recovery of specific pathogens _in situ_. However, in outbreaks where an etiological agent is not recovered, an epidemiological investigation may still implicate water as the vehicle of transmission. Whether or not an individual becomes ill depends on a series of complex interrelationships between the host and the infectious agent. Specific variables include (1) numbers of the invading microorganism (infectious dose), (2) the organism's ability to cause disease (pathogenicity), (3) the degree to which the microorganism can cause disease (virulence), and (4) the relative susceptibility of the host. In general terms, infectious microorganisms have the _potential_ to cause disease _only if_ the organism is sufficiently virulent to overcome the host's defenses.

Infectious Agents in Raw Domestic Wastewater

Any attempt to catalog the infectious agents or microbial pathogens that could conceivably be present in raw domestic wastewater would result in a surprisingly long list. The tremendous variety of infectious agents that might be present is derived principally from the feces and other body discharges of infected human and animal hosts. The occurrence and relative density of these agents in a

TABLE 3-8 Compounds Identified in Water Factory 21 Influent and Effluent[a,b]

Aromatic Hydrocarbons	Synthetic Chlorinated Compounds	Chlorination Products
Benzene	Methylene chloride[c]	Chloroform[c]
Toluene[c]	Trichloroethylene[c]	Dichlorobromomethane[c]
Ethyl benzene[c]	1,1,1-Tetrachloroethylene[c]	Chlorodibromomethane[c]
p-Xylene[c]	1,1,1-Trichloroethane[c]	Bromoform[c]
m-Xylene[c]	1,1,2-Trichloroethane[c]	Dichloroiodomethane[c]
o-Xylene[c]	Hexachloroethane[c]	1,1,2,2,-Tetrachloroethane[c]
1-Ethyl-4-methylbenzene[c]	Chlorobenzene[c]	β-Chlorostyrene isomers[c]
1-Ethyl-3-methylbenzene[c]	1,2-Dichlorobenzene[c]	Chlorobromoiodomethane[c]
1,3,5-Trimethylbenzene[c]	1,3-Dichlorobenzene[c]	1,1,1-Trichloroacetone[c]
1-Ethyl-2-methylbenzene[c]	1,4-Dichlorobenzene[c]	Chloroxylene[c]
1,2,4-Trimethylbenzene[c]	1,3,5-Trichlorobenzene[c]	Chlorobromopentanone[c]
1,2,3-Trimethylbenzene[c]	1,2,4-Trichlorobenzene[c]	Bromoketone[c]
C$_4$-Benzenes	PCB (Aroclor 1242)[c]	Methylchlorobenzene[c]
Lindane	Pentachlorophenylmethylether	α,β-Dichloroethylbenzene
Methylindanes	Tetrachlorophenylmethylether	
Naphthalene[c]	isomers	
1-Methylnaphthalene[c]	Trichlorophenylmethylether	
2-Methylnaphthalene[c]	Dichlorophenylmethylether	
C$_2$-Naphthalenes	Carbon tetrachloride[c]	
C$_3$-Naphthalenes	Lindane	
Styrene	Tetrachlorobenzene isomer	
Biphenyl		
Phenanthrene/anthracene		
Methylphenanthrene (four isomers)[c]		
Phenylnonane isomers		
Phenylundecane isomers		
C$_3$-Biphenyl isomers		
C$_6$-Biphenyl isomers		
Pyrene/fluoranthene		

Natural Products	Phthalate Esters	Miscellaneous Compounds
Terpenes	Dimethylphthalate[c]	Camphor
Terpene alcohols	Diethylphthalate	Isophorone
Fenchone	Di-n-butylphthalate[c]	p-tert-Amylphenol
Fenchyl alcohol	Diisobutylphthalate[c]	Octylcyanide
trans-β-Farnesane	Bis(2-ethylhexyl)phthalate[c]	Hexylcyanide
Heptaldehyde		Other alkyl cyanides
Lauric acid methyl ester[d]		Methyl-2-(p-chlorophenoxy)-
Myristic acid methyl ester[d]		2-methyl-propionate[c]
Peritadecanoic acid methyl		o-Isomer clofibrate metabolite
ester isomers[d]		Methylbenzoate
Heptadecanoic acid methyl		Tolualdehyde isomers[c]
ester[d]		Methylphenol
Stearic acid methyl ester		2-Chloropyridine
Palmitic acid methyl ester		Benzaldehyde[c]
		Pentadecane[c]
		Octadecane[c]
		Acetophenone[c]

[a]From McCarty et al., 1980. Priority pollutants are underlined.
[b]October 1977-December 1978 study where activated sludge effluents were split, treated by chlorination, and RO.
[c]Found in effluents.
[d]Identified after methylation.

TABLE 3-9 Concentrations of Organic Contaminants and Effectiveness of Treatment at Water Factory 21 in Their Removal[a]

Contaminant	Concentration (μg/liter)			Percent Removal		Dominant Removal Processes[c]
	AWT[b] Plant Influent	AWT[b] Plant Effluent	RO Effluent	AWT[b]	AWT[b] & RO	
Aromatic hydrocarbons						
Ethylbenzene	0.043	0.014	0.019	67	56	NR[d]
m-Xylene	0.035	0.020	0.024	43	31	NR
p-Xylene	0.015	0.012	0.014	20	7	NR
Naphthalene	0.033	0.010	0.028	70	15	NR
1-Methylnaphthalene	0.008	0.009	0.00	-12	88	NR
2-Methylnaphthalene	0.010	0.02	0.008		20	NR
Styrene	0.048	0.003		94		NR
Synthetic chlorinated compounds						
Carbon tetrachloride	0.033	0.16	0.008	-380	76	Strip
1,1,1-Trichloroethane	3.25	0.20	0.083	94	97	Strip
Trichloroethylene	0.74	0.10	0.1	86	86	Strip
Tetrachloroethylene	1.67	0.83	0.20	50	88	Strip
Chlorobenzene	0.14	0.049	0.034	65	76	Strip, GAC

Compound						Process
1,2-Dichlorobenzene	0.64	0.02	0.001	97	99.9	Strip, GAC
1,3-Dichlorobenzene	0.16	0.02	0.004	88	97.5	Strip, GAC
1,4-Dichlorobenzene	1.85	0.012	0.015	99	99	Strip, GAC
1,2,4-Trichlorbenzene	0.11	0.02	0.02	82	82	Strip, GAC
Lindane	0.14	0.05	0.05	64	64	ppt, GAC
PCB (Aroclor 1242)	0.47	0.3	0.3	36	36	ppt, GAC
Chlorination products						
Chloroform	3.1	8.6	0.97	-177	69	Strip, GAC
Dichlorobromomethane	0.53	2.7	0.24	-410	55	Strip, GAC
Chlorodibromomethane	0.69	1.3	0.13	-88	81	Strip, GAC
Bromoform	0.40	0.38	0.007	5	98	Strip, GAC
Phthalate esters						
Dimethylphthalate	4.8	0.47	1.0	90	79	ppt, GAC
Diethylphthalate	0.097	0.3	0.3			ppt
Di-n-butylphthalate	0.79	0.33	1.1	58	-39	ppt, GAC
Diisobutylphthalate	4.7	0.27	0.23	94	95	ppt, GAC
Bis-(2-ethylhexyl) phthalate	11.0	3.1	2.9	72	74	ppt

[a] From McCarty et al., 1980.
[b] AWT refers to all advanced wastewater treatment processes except RO.
[c] Strip = chemical precipitation with time
GAC = granular activated carbon
ppt = chemical precipitation with lime
[d] NR = not reported.

TABLE 3-10 Efficacy of a Pilot Plant to Remove Toxic Organic Compounds[a]

Compound	Concentration in Feedwater (µg/liter)	Percentage Removal by Unit Process (B) and Percentage Overall Removal (C)									
		High Lime Treatment		Secondary Clarification		Sand Filtration		Chlorination		Active Carbon Treatment	
		B	C	B	C	B	C	B	C	B	C
Lindane	20	0	0	0	0	0	0	80	80	98.3	99.5
Dieldrin	40	12.5	12.5	0	12	22	30	4	33	99	99.3
Chlordane	300	18	18	2	20	13	30	32	52	99	>99.5
Demeton-S-methyl	3,000	67	67	0	67	21	74	99.6	>100	--	--
Parathion	4,000	19	19	0	19	18	33	99.9	100	--	--
Fenitrothion	4,000	22	22	0	22	21	38	99.9	100	--	--

Compound	Conc									
Fenthion	4,000	22	22	0	22	19	37	58	74	100	100
Phenol	600	5	5	30	33	0	33	82	88	98	99.7
Hexachlorobutadiene	100	72	72	97	>99	--	--	--	100	--	
Acenaphthene	600	27	27	95	97	13	98	75	98		
Fluoranthene	500	79	79	80	96	33	98	22	100		
Pyrene	500	85	85	86	98	20	99	50	100		
Dibutylphthalate	400	87	87	0	53[b]	0	35[b]	28	53	99	100
o-Nitrotoluene	400	24	24	73	80	10	82	18	85	93	100
Tetradecane	200	96	96	89	100						

[a]From Van Rensburg et al., 1980.
[b]Increase in dibutylphthalate probably due to plastic piping used in pilot plant.

given raw domestic wastewater depends on a number of complex factors; therefore, it is impossible to define the general characteristics of a particular wastewater with respect to infectious agents.

The principal infectious agents in raw domestic wastewater can be classified within four broad groups: bacteria, viruses, protozoa, and helminths. Table 3-11 lists some of the infectious agents, along with their associated diseases. This list is representative--not all-inclusive; furthermore, the infectious agents listed have not been detected with the same degree of frequency in all raw domestic wastewaters. Although the absolute density cannot be given with any degree of accuracy, the variation and/or the order of magnitude of the density of certain infectious agents that might be encountered in raw domestic wastewater can be illustrated, e.g., Salmonella, up to 10^4/liter; protozoa (cysts), up to 10^5/liter; helminths (ova), up to 10^3/liter; enteric virus plaque-forming units (pfu), up to 10^5/liter. Such information should be considered with respect to the inadequacies of the detection and enumeration methods used for some infectious agents. For example, the observed density of enteric viruses may be 1 to 2 logs lower than the actual density due to the limitations of the current virus recovery and cultivation procedures. Notwithstanding these limitations, it is accepted that these infectious agents can be found in raw domestic wastewater and, on occasion, in a water supply.

Bacteria

The most common infectious agents in raw domestic wastewater are perhaps the enteric bacteria; of these, members of the genus Salmonella apparently occur most frequently. Salmonella have frequently been isolated from feces, wastewater, receiving water, and occasionally from finished water supplies. Shigella organisms are among the chief etiological agents of bacillary dysentery. There are few reports of Shigella being detected in wastewater, even though Salmonella may be routinely isolated. Survival of Shigella in wastewater is relatively short. The possible presence of enteropathogenic Escherichia coli in raw domestic wastewater is also of concern. These agents were first associated with outbreaks of diarrhea in nurseries (Craun, 1978). The prevalence of enteropathogenic E. coli infections in the United States is uncertain, but this infective agent can be involved in waterborne outbreaks of gastroenteritis. Certain strains of E. coli can also cause gastroenteritis through the production of enterotoxin in the small intestine. In addition to these three genera of pathogenic bacteria, others may occur, but less frequently, in raw domestic wastewater. Among these are species of Vibrio, Clostridium, Leptospira, Mycobacterium, Campylobacter, and Yersinia.

Viruses

More than 100 distinct serotypes of viruses have been identified in raw domestic wastewater (Table 3-12), and the density in the United

TABLE 3-11 Infectious Agents Potentially Present in Raw Domestic Wastewater

Organism	Disease
BACTERIA	
Shigella (4 spp.)	Shigellosis (bacillary dysentery)
Salmonella typhi	Typhoid fever
Salmonella (1,700 spp.)	Salmonellosis
Vibrio cholerae	Cholera
Escherichia coli	Gastroenteritis
Yersinia enterocolitica	Yersinosis
Leptospira (spp.)	Leptospirosis
Campylobacter	Gastroenteritis
VIRUSES	
Enteroviruses (71 types)	Gastroenteritis, heart anomalies, meningitis
Hepatitis A virus	Infectious hepatitis
Adenovirus (31 types)	Respiratory disease
Rotavirus	Gastroenteritis
Reovirus	Not clearly established
Gastroenteritis virus (Norwalk-type)	Gastroenteritis
PROTOZOA	
Endamoeba histolytica	Amebiasis (amoebic dysentery)
Giardia lamblia	Giardiasis
Balantidium coli	Balantidiasis (balantidial dysentery)
HELMINTHS	
Ascaris lumbricoides	Ascariasis
Ancylostoma duodenale	Ancylostomiasis
Necator americanus	Necatoriasis
Ancylostoma (spp.)	Hookworm
Strongyloides stereoralis	Strongyloidiasis
Trichuris trichiura	Trichuriasis
Taenia (spp.)	Taeniasis
Enterobius vermicularis	Enterobiasis
Echinoccoccus granulesis	Hydatidosis
Schistosoma mansonii	Schistosomiasis

TABLE 3-12 Enteric Viruses Potentially Present in Water[a]

Virus Group	No. of Types	Disease Caused
Enteroviruses		
Poliovirus	3	Paralysis, meningitis, fever
Echovirus	34	Meningitis, respiratory disease, rash, diarrhea, fever
Coxsackievirus A	24	Herpangina, respiratory disease, meningitis, fever
Coxsackievirus B	6	Myocarditis, congenital heart meningitis, anomalies, rash, fever, respiratory disease, pleurodynia
New enteroviruses	4	Meningitis, encephalitis, respiratory disease, acute hemorrhagic conjunctivitis, fever
Hepatitis type A	1	Infectious hepatitis
Gastroenteritis virus	?	Epidemic vomiting and diarrhea, fever
Rotavirus	?	Epidemic vomiting and diarrhea, chiefly of children
Reovirus	3	Not clearly established
Adenovirus	30	Respiratory disease, eye infections

[a]Adapted from World Health Organization, 1979.

States may be on the order of 10^4 viral units/liter (World Health Organization, 1979). Enteric viruses are those that multiply in the intestinal tract and are shed in the feces of infected persons. Of the many enteric viruses detected in wastewater, those associated with infectious hepatitis and gastroenteritis (Norwalk-type) warrant special attention because of the existing epidemiological evidence that they can be spread via the water route. Unfortunately, these viral agents have yet to be cultivated in the laboratory; thus, no recovery method is yet available to determine their density in wastewater and water supplies. Although recently improved techniques have increased the sensitivity of virus recovery, resulting in higher numbers in water than those previously reported, there is still a need for a better understanding of the reliability, limits of detection, and precision of the methodology for recovery and culture of enteric viruses. Also, today's procedures require a 2-week period before a water sample can be recorded as negative for enteric viruses. There is also the problem of detecting and enumerating waterborne viruses that are attached or embedded in particulate matter. The significance of virus clumping or aggregation with respect to virus recovery and enumeration in water, and the interpretation of such information in terms of infectious unit or dose, also need further examination.

Protozoa

The protozoa of greatest concern because of disease transmission through water are Endamoeba histolytica, Giardia lamblia, and, to a lesser extent, Balantidium coli. The reservoir of infection for E. histolytica is in humans. Although amebiasis is rare in the United States, carrier-type infections can be found among the population. These carriers may shed a large quantity of cysts, e.g., 10^7/day (Jukubowski and Ericksen, 1979), sufficient for their detection in raw domestic wastewater. In recent years, the protozoan of greatest health concern has been the flagellate G. lamblia. Giardiasis appears to be endemic in the United States as Giardia is frequently identified in stool samples; as many as 10^8 cysts may be excreted daily by infected individuals. The potential exists, therefore, for the disease to be spread through water contaminated with fecal matter.

Helminths

There are some two dozen helminths, belonging to either of two phyla--Platyhelminthes (flatworms) or Nematoda (roundworms)--which may cause infections in humans by transmission through water. Improved sanitary conditions have greatly reduced the prevalence of helminth infections among human populations. However, water transmission of such helminthic infections as ascariasis and trichuriasis is still possible when the infective stage of the helminth is ingested with water. Other helminths gain entrance to humans by way of skin penetration, e.g., hookworms (Necator and Ancylostoma) and schistosomes (Schistosoma); the presence of these helminths in contaminated water can be transmitted to humans during bathing. With some helminths, the infective stage occurs during either the adult or a larval form of the organism; with the others (and more commonly), the ova are important in the spread of the infection through water. The ova of intestinal parasitic worms are excreted in the feces of infected individuals; consequently, they may be found in raw domestic wastewater. Indeed, the ova of Ascaris lumbricoides, the whipworm Trichuris trichiura, the tapeworm Taenia saginata, and others have been observed in wastewater.

In addition to these, a variety of other infectious agents may be present in raw domestic wastewater, e.g., infectious yeasts and molds, viroids, and rickettsia. Evaluation of the presence of these and perhaps others of health significance appears to have been neglected. The presence and significance of antibiotic-resistant bacteria in the water environment also need further attention. This phenomenon is associated with nucleic acid elements or resistance (R) factors capable of rapid transfer by conjugation between gram-negative bacteria.

Among the substances of physiological concern that might be present in wastewater are microbial toxins that can produce a physiological response after ingestion. They have been associated with species of <u>Escherichia</u>, <u>Vibrio</u>, <u>Clostridium</u>, and <u>Staphylococcus</u>, among others. The production of exotoxins may be associated with the carryover of excessive amounts of organic suspended solids from biological wastewater treatment processes and the subsequent decomposition of these materials by anaerobic bacteria in water storage tanks.

Removal and/or Inactivation of Infectious Agents

The initial treatment of raw domestic wastewater for water reuse will probably continue to involve those processes normally included in conventional wastewater treatment, i.e., primary treatment or sedimentation and biological secondary treatment with final clarification. In addition to preparing the wastewater for disinfection, it is generally recognized that primary and secondary treatment of raw domestic wastewater removes such infectious agents as bacteria, viruses, protozoan cysts, and helminth ova, but that the success of the removal is variable at best (Table 3-13), depending in part on the nature of the infectious agents (Engelbrecht and Lund, 1975). Primary sedimentation may remove up to 50% of the bacterial population, including pathogenic bacteria, in raw domestic wastewater. Although there is insufficient information, it appears that, because of a higher specific gravity, helminth ova are more effectively removed by sedimentation than are amoebic cysts. Only a small percentage of the enteric viruses present in raw domestic wastewater appear to be removed by primary treatment. However, viruses are known to adsorb to particulate matter and probably occur as individual particles only on rare occasions. Viruses may also be embedded in particulate matter. These factors lead one

TABLE 3-13 Removal of Representative Infectious Agents by Conventional Wastewater Treatment[a]

Infectious Agents	Primary Treatment (% Removed)	Secondary Treatment (% Removed)	
		Activated Sludge	Trickling Filter
<u>Salmonella</u>	15	90-99	90-99.9
<u>Mycobacterium</u>	40-60	5-90	70-99
<u>Shigella</u>	15	80-90	85-99
Amoebic cysts	Limited	Limited	10-99.9
Helminth ova	70-95	Limited	60-75
Viruses	Limited	75-99	0-86

[a]From Engelbrecht and Lund, 1975.

to suspect that virus removal by sedimentation has not been accurately measured and that more viruses are actually removed than the data indicate. The same may be true of other infectious agents.

As with primary treatment, the removal of infectious agents by secondary treatment, i.e., activated sludge or trickling-filter followed by final clarification, is variable (Table 3-13). Bacteria removal may be as high as 90%. Amoebic cysts and helminth ova appear to be much more effectively removed by trickling-filter treatment than by activated sludge. Of these two secondary biological processes, activated sludge appears to achieve the greatest and most consistent removal of viruses. Of course, numerous microorganisms, including bacteria, fungi, and algae, grow in biological wastewater treatment processes, and some of these may be present in the effluent. Little is known about the health significance of these microorganisms or of their products, or of their fate in subsequent treatment processes directed toward the potable reuse of water.

From Table 3-13 and this brief discussion, it is apparent that conventional wastewater treatment, consisting of primary and secondary treatment, reduces the density of infectious agents in wastewater, but does not produce an effluent free of infectious agents. On the contrary, a significant number of viable infectious agents may be present in secondary effluent. The microbiological quality of a secondary effluent can be further improved, however, through disinfection. The enteric bacterial infectious agents may be effectively inactivated by disinfection using chlorine. Although data are lacking, protozoan cysts and helminth ova, because of their general resistance to disinfection, might be ineffectively inactivated by chlorination of a secondary effluent as currently practiced. Furthermore, certain enteric viruses appear to be more resistant to chlorine than are coliforms and, perhaps, other bacteria. Ozonation of a secondary effluent can also be effective in deactivating infectious agents, giving results equal to or better than those produced by chlorination. In designing any disinfection system, however, it is important to consider the inadvertent formation of toxic by-products.

The treatment of secondary effluent by processes more commonly associated with water purification can further reduce the density of infectious agents. These processes might include chemical coagulation, filtration, activated carbon adsorption, and disinfection. Chemical coagulation-flocculation, particularly with lime at high pH (>11), followed by clarification appears to be highly effective in removing enteric bacteria and viruses (Engelbrecht, 1976). Although there seem to be no data on high pH treatment, it is reasonable to assume that protozoan cysts and helminth ova and other stages in the life cycle of these parasites would also be removed from water by such treatment followed by clarification.

By itself, rapid sand filtration is not effective in removing viruses and, perhaps, bacteria. However, if preceded by chemical coagulation and by clarification, rapid sand filtration is effective (Engelbrecht, 1976). Porous media filtration, following coagulation and/or the addition of polymer, also appears to be effective in removing protozoan cysts (including <u>Giardia</u>) and helminth ova. In addition to removing infectious agents, filtration reduces turbidity--a step that is essential for effective disinfection.

The final stage of treatment in any water reclamation scheme is terminal disinfection. Based on existing information, experience, and other considerations, the disinfectants of choice are either chlorine and/or ozone; another possibility is chlorine dioxide. Given the proper conditions, breakpoint or free residual chlorination (0.5 mg/liter for 60 min contact) has been reported to achieve a significant virus reduction (Hattingh, 1978). The reduction of coliform organisms, and presumably the pathogenic bacteria, would be similar under the same conditions. Although information is limited, protozoan cysts and helminth ova have been shown to be more resistant to chlorine than are bacteria and viruses. The effectiveness of ozone in deactivating bacteria and viruses is equal to or perhaps better than that of chlorine. There appears to be no information on the inactivation of helminths by ozone; that on protozoan cysts is limited. Data indicate that chlorine dioxide may be just as effective in inactivating enteric bacteria and viruses as is chlorine, but there is a lack of information on the efficacy of chlorine dioxide for the inactivation of protozoan cysts and helminths.

Anticipating areas of future acute water shortage in South Africa, the National Institute for Water Research of South Africa initiated an aggressive water reclamation research program in the early 1960's. This program included laboratory studies, as well as pilot and full-scale treatment plant investigations of various reclamation unit processes. As a result, a vast amount of information has been collected on the removal and/or inactivation of infectious agents in secondary wastewater effluent by chemical coagulation-flocculation, filtration, activated carbon adsorption, disinfection, and other treatment processes. This information has been summarized by the National Institute for Water Research and is included in Table 3-14 (Hattingh, 1978). Performance of these treatment plants indicates that, from a microbiological point of view, major attention has been given to the removal and/or inactivation of enteric bacteria and viruses; in most cases, the protozoan cysts and helminths have not been considered to any great extent.

The experimental water reclamation plants at Windhoek, Namibia, (formerly South-West Africa), and at Pretoria (Stander Water Reclamation Plant), South Africa (described in Chapter 2), have provided a substantial amount of information on the removal and/or inactivation of infectious agents after various degrees of treatment as well as of the final product water, particularly with respect to enteric viruses and bacteria (Grabow et al., 1978; Nupen et al., 1974). No enteric viruses were detected (using two different concentration procedures) in 286 samples representing a volume of 1,788 liters (152 10-liter samples and 134 2-liter samples) taken after the final four treatment stages of the Windhoek plant and 461 samples representing a volume of 4,610 liters (461 10-liter samples) obtained following the last three treatment units in the Stander plant (Grabow et al., 1978). Many more samples were taken for density of total bacteria, total coliforms, fecal coliforms, enterococci, Clostridium perfringens, Pseudomonas aeruginosa, Staphylococcus aureus, and Escherichia coli B coliphage. During normal operation, the final product water at both the Windhoek and Stander plants was reported to meet the following limits: total

TABLE 3-14 Reduction of Fecal Coliform and Viruses by Various Reclamation Unit Processes[a]

Unit Processes	Log Reduction[b]	
	Fecal Coliforms	Virus
Coagulation--aluminum sulfate or ferric chloride	1-2	1-2
Lime treatment, pH 11.2	2	3-4
Lime treatment, pH 11.5	5	4-5
Activated carbon adsorption	0.5	5
Rapid sand filtration	1-2	1-2
Chlorination, <breakpoint	11	
Chlorination, >breakpoint, low turbidity, and nitrogen	7	7

[a]From Hattingh, 1978.
[b]Results achievable under favorable operating conditions.

bacterial plate count, 100/1 ml; other bacteria, 0/100 ml; enteric viruses, 0/10 liter; and coliphage, 0/10 ml.

The use of acid-fast bacteria as an indicator group for assessing the microbiological quality of reclaimed wastewater has also been evaluated (Grabow et al., 1981). These bacteria were found to be exceptionally resistant to most treatment processes. The authors concluded that their absence after chlorination ensured the absence of vegetative bacteria and enteric viruses. The total plate count was found to be a highly sensitive indicator of any microbiological contamination, and the coliphage proved found to be a rapid, economical, and simple method for screening removal and/or inactivation of viruses. Monitoring for parasitic ova has also been performed at both the Windhoek and the Stander reclamation plants by direct microscopic examination of 10 liters of membrane-filtered concentrate (Grabow and Isaacson, 1978). No parasitic ova were detected in the final product water.

The Environmental Protection Agency performed an 18-month advanced wastewater treatment study at its Blue Plains pilot plant in Washington, D.C., to obtain data on the effluent discharged upstream of drinking water intakes in order to determine its safety for potential domestic reuse purposes (Warner et al., 1978). Samples for detecting enteric viruses were processed during three different periods of operation. The density of viruses in the raw wastewater ranged from 1,850 to 18,000 pfu/100 liters during the three sampling periods. Viruses could not be detected following filtration during the first sampling period, in the effluent from the biological nitrification unit during the second sampling period, or following chlorination in the samples taken during all three sampling periods. During the entire 18-month study period, the average concentration of free available chlorine in the final effluent was 2.23 mg/liter (arithmetic mean of 57 determinations). The arithmetic mean densities of total coliforms (114 samples), fecal coliforms (119 samples), Pseudomonas aeruginosa (36 samples), Salmonella (40 samples), and total bacteria plate count (44 samples) following terminal chlorination were 0.11, 0.17, 0.81, 0, and 66.8/100 ml, respectively (Warner et al., 1978).

Water Factory 21 (described in Chapter 2) was designed to reclaim unchlorinated secondary wastewater effluent (McCarty et al., 1978, 1980). During one period of operation, unchlorinated trickling-filter effluent was treated by lime clarification (pH 11.3), ammonia stripping, recarbonation, mixed-media filtration, activated carbon adsorption, and postchlorination. Breakpoint chlorination for nitrogen control was also evaluated at different locations in the treatment sequence. During this period of operation, the geometric mean density of total coliforms and fecal coliforms in the unchlorinated trickling-filter effluent was 89×10^6 and 25×10^6 most probable number (MPN) per 100 ml, respectively. The reduction in total and fecal coliform bacteria by the lime clarification stage resulted in a 98% to 99.9% reduction in viruses, depending on the detection procedure used. Based on unconfirmed virus analysis, 48 of 77 trickling-filter effluent samples were found to be positive for viruses; the density, calculated as the geometric mean of the positive samples, was 1.1 pfu/liter. More than 25 different enteric virus types were identified in the trickling-filter effluent. Of 77 chlorinated, final effluent samples assayed, 2 samples were positive for viruses. Both positive samples were believed to be due to the presence of a high concentration of activated carbon fines in the samples, which may have interfered with proper postdisinfection.

In a subsequent period of operation, a high-quality activated sludge effluent was used as the influent to the advanced wastewater plant. The geometric mean density of total coliforms, fecal coliforms, and viruses in the activated sludge effluent during this period was 1.6×10^6 MPN/100 ml, 0.55×10^6 MPN/100 ml, and 0.13 most probable number of cytopathogenic units (MPNCU)/liter, respectively. The virus detection procedure used was judged to be about 3 times more sensitive than that used in the earlier study with trickling-filter effluent. Again, lime clarification removed more than 5 logs of the total coliform and fecal coliform bacteria. Following postchlorination, the geometric mean density of total coliforms and fecal coliforms was 0.05 and <1 MPN/100 ml, respectively; no viruses were detected in the final effluent, i.e., following postchlorination.

SUMMARY AND CONCLUSIONS

The relatively limited experience with actual wastewater treatment and pilot plant systems designed to give additional treatment to secondary municipal effluent for potable reuse has shown that the concentrations of chemical constituents in the final product do not usually occur at levels that are considered unacceptable in terms of current drinking water standards. Also, there is evidence that the sequence of treatment processes used or contemplated in such systems can remove a broad variety of types of inorganic and organic chemicals, some more effectively than others. However, a combination of processes is usually required to ensure effective treatment.

Although a few wastewater treatment systems are being studied extensively to determine the composition of the trace inorganic and organic constituents in their finished water, there is insufficient information to compare their water to a broad range of currently used potable water supplies and to judge whether either the latter or the treated waters are generally lower in trace chemicals content. This lack of data is especially true for organic chemicals, which have not been thoroughly characterized for most potable water.

Thus, although it may be argued that there is no evidence to rule out the direct use of treated wastewater for potable purposes (based on the limited data on trace chemical composition), it is also clear that the toxicological significance to human health of these and other, as yet unidentified chemicals has yet to be determined.

Although not unexpected, there is growing evidence of a wide variation in the efficacy with which the different chemical constituents are removed in such treatment systems and in the composition of the final treated water, depending on the wastewater source and the system treating it. In addition, within a given system there can be substantial variabilities among the chemical constituents over time. Because of such variabilities, and of differences among secondary wastewater effluents and wastewater treatment systems, it may be necessary to evaluate each potential treatment system individually with respect to its ability to remove chemical constituents effectively. At the same time, the constituents that vary most extensively over time may require more attention from both a toxicological point of view and from the design and monitoring effort.

A wide range of infectious agents may be found in raw domestic wastewater. The more usual infectious agents can be classified within four broad groups: bacteria, viruses, protozoa, and helminths. They are derived principally from infected persons and can be spread to others via the water. Other agents of concern include toxins, antibiotic-resistant bacteria, and fungi. Although some infectious agents cannot survive outside their host for any great length of time, their environmental transmission is normally controlled through wastewater and water treatment.

Today's technology appears capable of providing the degree of water treatment required to meet any microbiological criteria associated with using a severely contaminated raw water source such as domestic wastewater. Conventional wastewater treatment (consisting of primary and biological secondary treatment) may significantly reduce the density of infectious agents in raw domestic wastewater, but it does not produce an effluent free of infectious agents. Disinfection may improve the microbiological quality of a secondary effluent, particularly with respect to its bacterial population, but it is unlikely that the protozoan cysts, helminth ova, and enteric viruses possibly present in such effluent would be effectively inactivated by disinfection as practiced today. The density of infectious agents in secondary effluent can be reduced by the treatment processes more commonly associated with water purification; these include chemical coagulation-flocculation (particularly with lime at a pH >11), filtration, activated carbon adsorption, and terminal disinfection (e.g., breakpoint chlorination). When properly sequenced in a treatment train and under optimum operating conditions, these processes can produce a final product of acceptable microbiological quality. However, this observation does not eliminate the need for additional confirmatory data on the removal and/or inactivation of infectious agents by the various treatment processes, nor of the need to ensure the operational integrity of treatment systems. In this respect, reliable monitoring techniques must be developed for rigorous applications in any potable reuse scheme.

REFERENCES

Battelle Columbus Laboratory. 1980. GC/MS Analysis of Organics in Drinking Water, Water Concentration and Advanced Water Treatment Concentration. EPA Project No. 68-03-2548. Battelle Laboratory, Columbus, Ohio.

Chang, A.C., and A.L. Page. 1980. Fate of inorganic microcontaminants during groundwater recharge. Pp. 118-136 in Wastewater Reuse Groundwater Recharge. California State Water Resources Control Board, Sacramento, Calif.

Craun, G.F. 1978. Impact of the coliform standard on the transmission of disease. In C.W. Hendricks, ed. Evaluation of the Microbiology Standards for Drinking Water. EPA 570/9-78-00C. Office of Drinking Water, Environmental Protection Agency, Washington, D.C.

Engelbrecht, R.S. 1976. Removal and inactivation of enteric viruses by wastewater and water treatment processes. P. 109 in Proceedings of Advanced Wastewater Treatment Seminar. Japan Research Group of Water Pollution, Tokyo, Japan.

Engelbrecht, R.S., and E. Lund. 1975. Biological properties of wastewater sludge and its potential health risk. Conference Proceedings, International Water Conservancy Exhibition. Jonkoping, Sweden.

Englande, A.J., Jr., and R.S. Reimers III. 1979. Wastewater reuse-persistence of chemical pollutants. Pp. 1368-1389 in Proceedings, Water Reuse Symposium. American Water Works Association Research Foundation, Denver, Colo.

Environmental Protection Agency. 1976. National Interim Primary Drinking Water Regulations. EPA-570-9-76-003. Environmental Protection Agency, Washington, D.C.

Environmental Protection Agency. 1977. Manual of Treatment Techniques for Meeting the Interim Primary Drinking Water Regulations. EPA-600/8-77-005. Environmental Protection Agency, Washington, D.C.

Environmental Protection Agency. 1980. Concentration Degree Survey for Priority Pollutants in Industrial Effluents. Project I--Organic Analysis of Seven Municipal Wastewater Treatment Plants. Project II--Analysis of GC/MS Data Collected During Project I. Effluent Guidelines Division, Environmental Protection Agency, Washington, D.C.

Garrison, A.W., J.D. Pope, A.L. Alford, and C.K. Doll. 1979. An automatic sampler, a master analytical scheme and registry system for organics in water. In Trace Organic Analysis: A New Frontier in Analytical Chemistry. Special Publication 519. National Bureau of Standards, Washington, D.C.

Grabow, W.O.K., and M. Isaacson. 1978. Microbiological quality and epidemiological aspects of reclaimed water. Progress Water Technol. 10(1-2):329.

Grabow, W.O.K., B.W. Bateman, and J.S. Burger. 1978. Microbiological quality indicators for routine monitoring of wastewater reclamation systems. Progress Water Technol. 10(5):317.

Grabow, W.O.K., H.S. Burger, and E.M. Nupen. 1981. Evaluation of acid-fast bacteria, Candida albicans, enteric viruses and conventional indicators for monitoring wastewater reclamation systems. Progress Water Technol. 13(2):803.

Hattingh, W.H.J. 1978. Health aspects. P. 257 in Manual for Water Renovation and Reclamation. Technical Guide K42. Council for Scientific and Industrial Research, Pretoria, South Africa.

Hrubec, J., J.C. Schippers, and B.J. Zoetman. 1979. Studies on water reuse in The Netherlands. Pp. 785-807 in Proceedings, Water Reuse Symposium. American Water Works Association Research Foundation, Denver, Colo.

Jukubowski, W., and T.H. Ericksen. 1979. Methods for detection of Giardia cysts in water supplies. Pp. 193-210 in W. Hakubowski and J.C. Hoff, eds. Proceedings of Symposium on Waterborne Transmission of Giardiasis. EPA-600/9-79-001. Environmental Protection Agency, Cincinnati, Ohio.

McCarty, P.L., M. Reinhard, C. Dolce, H. Nguyen, and D.G. Argo. 1978. Water Factory 21: Reclaimed Water, Volatile Organics, Virus, and Treatment Performance. EPA 600/2-78-076. Municipal Environmental Research Laboratory, Office of Research and Development, U.S. Environmental Protection Agency, Cincinnati, Ohio. 87 pp.

McCarty, P.L., M. Reinhard, J. Graydon, J. Schreiner, K. Sutherland, T. Everhart, and D.G. Argo. 1980. Advanced Treatment for Wastewater Reclamation at Water Factory 21. Technical Report 236. Department of Civil Engineering, Stanford University, Stanford, Calif. 149 pp.

Moss, C.E. 1973. Control of radioisotope releases to the environment from diagnostic isotope procedures. Health Phys. 25:197-198.

National Academy of Sciences. 1977. Drinking Water and Health. Safe Drinking Water Committee, Advisory Center on Toxicology, Assembly of Life Sciences, National Research Council. National Academy of Sciences, Washington, D.C. 939 pp.

Nupen, E.M., B.W. Bateman, and N.C. McKenny. 1974. The reduction of virus by the various unit processes used in the reclamation of sewage to potable water. In J.F. Malina and B.P. Sagik, eds. Virus Survival in Water and Wastewater Systems. Water Resources Symposium No. 7. Center for Research in Water Resources, University of Texas at Austin, Austin, Tex.

Shackelford, W.M., and L.H. Keith. 1976. Frequency of organic compounds identified in water. Environmental Research Laboratory, U.S. Environmental Protection Agency, Athens, Ga. 618 pp.

Stander, G.J. 1980. Micro-organic compounds in the water environment and their impact on the quality of potable water supplies. Water SA 6(1):1-14.

Van Rensburg, J.F.J., A.J. Hassett, and S.J. Theron. 1980. Control and measurement of organic micropollutants in South African water reclamation plants. Presented at Symposium on Chemistry and Chemical Analysis of Water/Wastewater Intended for Reuse, Houston, Tex. American Chemical Society, Washington, D.C.

Warner, H.P., J.N. English, and I.J. Kugelman. 1978. Wastewater treatment for reuse and its contribution to water supplies. P. 36 in USA-USSR Working Group on the Prevention of Water Pollution from Municipal and Industrial Sources: Symposium on Advanced Equipment and Facilities for Wastewater Treatment. Office of Water Programs Operations, Environmental Protection Agency, Washington, D.C.

World Health Organization. 1979. Human Viruses in Water, Wastewater and Soil. Technical Report Series No. 639. World Health Organization, Geneva, Switzerland.

4

Concentration Methods for Analysis and Toxicity Testing[1]

Before trace amounts of constituents in aqueous solutions as complex mixtures can be chemically analyzed, the solutions must be concentrated. This step is necessary so that a sufficient mass of chemicals can be obtained for separation and subsequent identification. An analogous situation exists for determining the toxicity of such unknown or mostly unknown trace constituents in waters of environmental or public health concern.

The choice of method or combination of methods for concentration depends on such factors as volatility of the constituent to be tested, the degree of concentration required, and the toxicological test system to be used.

As reported by Jolley (1980), Kopfler (1980) divides concentration methods into two basic categories:

1. Concentration: That is, those processes in which water is removed and the dissolved substances are left behind. Examples are freeze concentration, lyophilization (freeze drying), vacuum distillation, and membrane processes such as reverse osmosis and ultrafiltration. A common disadvantage to these methods is that inorganic species are concentrated along with the organic constituents.
2. Isolation: That is, those processes in which the chemicals are removed from the water. Examples are solvent extraction and adsorption on resins.

This chapter focuses primarily on trace organic chemicals, a major concern in wastewater treatment systems. However, many of the principles and concerns apply to inorganic constituents as well. The utility of concentration techniques for subsequent bioassay and toxicity testing is also discussed; many of these same considerations are also relevant to analyzing and monitoring reuse systems.

Although a variety of methods have been reported for concentrating organic (and inorganic) solutes in water, the following sections are confined to reverse osmosis and macroreticular resins. At present, these procedures hold the most promise for concentrating the greatest

---

[1]See Appendix A for further details.

fraction of the total organic carbon (TOC) in water and wastewater samples, i.e., the nonvolatile, relatively polar, and/or large molecular weight fractions.

REVERSE OSMOSIS

Hindin et al. (1969) used reverse osmosis (RO) with a cellulose acetate membrane for concentrating more than 30 organic compounds in water. Rejection by the membrane was found to depend on particle size, ionization, and volatility in aqueous solution. Compounds with a high vapor pressure in water may permeate the membrane. Degree of hydration was also important for amino acids. When mixtures of compounds were tested, rejection increased for some species and decreased for others.

Attempts have been made to predict rejection of a particular solute by a given membrane from solubility parameters (Deinzer et al., 1975; Klein et al., 1975). In some cases, prediction was poor because of adsorption of the solute or obvious interaction with the membrane. Klein et al., (1975) suggested using complementary membranes (based on their solubility coordinates), which would retain different types of solutes. Because ionizable solutes and polyhydroxy compounds are rejected by cellulose triacetate membranes, and aromatic compounds are expected to be rejected with greater efficiency by a nylon membrane, a combination of membranes should increase recovery for a mixture of organic constituents.

Kopfler et al. (1977) used such a two-step RO system with a cellulose acetate membrane and a nylon membrane for concentrating organic compounds in drinking water samples. In addition, a Donnan softening loop was used to reduce calcium and magnesium concentrations for ion exchange with sodium. The use of the softening unit did not reduce the yield of organic compounds recovered, but it did extend the life of the membrane by retarding scaling and membrane fouling. The concentrates obtained from the RO procedure were further concentrated by solvent extraction and adsorption on XAD-2 resin. Approximately 20% of each fraction was used for chemical analysis, and the remaining 80% was dried at room temperature for later toxicity studies. Recovery efficiency was high for the cellulose acetate membrane, but could not be calculated for the nylon membrane because the TOC in the influent was too low. The average estimated recovery for the entire process was 35.89%. A control experiment with distilled water indicated the presence of a number of organic compounds, 28 of which were identified by gas chromatography (GC) and mass spectrometry (MS). It was not determined if the contaminants came from the distilled water or from the membranes.

Fang and Chian (1976) examined 12 RO membranes for concentrating 13 polar, low molecular weight organic compounds in water, separately and mixed. The compounds contained a variety of functional groups. The cross-linked polyethylenimine membranes provided better separation (>75% TOC) of the low molecular weight, polar organic compounds tested than did the other membranes, including the cellulose acetate membranes. No quantitative relationship was found between solute separation and solubility parameters. Sewage effluent concentrated with an aromatic polyamide membrane yielded better recovery than that observed with the test compounds, an expected result because of higher molecular weight compounds in the sewage effluent. Also, membrane separation of TOC increased with the degree of treatment of the sewage effluent.

MACRORETICULAR RESINS

Most of the adsorbants used are either polystyrenedivinylbenzene (DVB) copolymers or polymethacrylate materials cross-linked with a suitable nonaromatic material (Gustafson and Paleos, 1971).

The important adsorptive properties of the resins include van der Waals' forces, dipole interactions, and hydrogen bonding (Thurman et al., 1978). In general, adsorption increases as the water solubility of the organic compound decreases. And, in a series of related compounds, adsorption via hydrophobic bonding increases as the length of a hydrocarbon chain increases, as the number of hydrocarbon substituents increases, or as the number of aromatic rings increases (Gustafson and Paleos, 1971).

A study using macroreticular resins XAD-2 and XAD-7 to concentrate model mixtures of organic compounds and contaminated well water showed that many nonionic organic compounds were extracted with about 100% efficiency, even at concentrations in the parts per billion range (Burnham et al., 1972). Common inorganic ions, such as Na^+ and Cl^-, as well as strongly ionic compounds (benzenesulfonic acid, p-toluenesulfonic acid, and 4-naphtholsulfonic acid) were not retained. The weakly acidic or basic organic compounds (carboxylic acids, phenols, and amines) were sorbed to varying degrees, depending on the pH of the solution.

The results of another extensive survey (Junk et al., 1974) indicated that sorption on XAD resins with an efficiency of extraction between 80% and 100% applies to a large number of compounds added to water.

There has been an effort to develop a model to determine which solutes adsorb and to ascertain their respective capacity factors (Thurman et al., 1978). The logarithm of the capacity factor k' was found to correlate inversely with the logarithm of the aqueous molar solubility S. Differences among k' measurements (20 solutes) conformed to the earlier results; that is, XAD-8 favored aliphatic over aromatic over alicyclic carbon systems (inverse solubility trend). Likewise, the following functional groups were preferred: $-CH_3 > -CO_2 > -CHO > -OH > NH_2$. Also, the solubility factor was empirically related to bed volume (Y, cm^3) and sample size (X, ml) in the following way: $X = 1/6\ k'y$. Thus, it is possible to match column and sample size in the preconcentration for efficient removal of organic solutes from water. Another consequence is that, given the amount of resin and sample used, the k' or solubility of solutes completely retained by the resin can be estimated.

For resins of similar chemical composition, the extent of adsorption has been reported to increase with increasing surface area (Gustafson and Paleos, 1971); however, in some cases no significant differences have been found between two low polarity resins (Junk et al., 1974).

A variety of adsorbents have been studied for their ability to adsorb polychlorinated biphenyls (PCB's) from synthetic aqueous solutions and raw sewage (Lawrence and Tosine, 1976). Amberlite XAD-2 was found to strongly adsorb the PCB's from aqueous solutions, but it was much less effective when used with raw sewage (efficiency approximately 23%). Amberlite XAD-4 was, after polyvinyl chloride (PVC), the second most effective adsorbent in raw sewage (about 60% and 73%, respectively).

BVP (poly[N-benzyl-4-vinylpyridinium]) resin (in bromide form) has been found to have a greater capacity for removing sodium dodecylbenzenesulfonate (DBS) from aqueous solutions, as opposed to commercial anion exchange resins (IRA-400 and IR-45) and porous styrenedivinylbenzene resins with no ion exchange functional groups (XAD-2 and XAD-4) (Kawabata and Morigaki, 1980).

In the isolation of organic impurities from water, there are several critical steps:

1. Resin purification and handling: The porous polymer resins are best cleaned by Soxhlet solvent extraction, and they must be kept in a continuously wetted condition to ensure a very low blank (Junk et al., 1974). Purity of resins varies (Fan et al., 1978).

2. Preparation of standard samples: If proper techniques are not used when adding organic impurities to water, errors in recovery are most probable whenever the concentration level exceeds the solubility of the organic compound in the water, the compound is less dense than the water and, thus, rises to the surface, and the volatility of the compound is appreciable (Junk et al., 1974).

3. Suppression of ionization: For organic acids, the preconcentration must be done at a pH two units below pK_a and for organic bases at a pH two units above pK_b (Thurman et al., 1978).

4. Concentration: Vessel shape has been proved critical, and free evaporation aided by a stream of nitrogen or other gas detrimental (Junk et al., 1974).

In addition, it has been shown that some of the compounds in the final concentrates are derived from the metal and plastic comprising the plumbing, membranes, etc., required by the process (Smith, 1978).

CONCENTRATION METHODS AND TOXICITY TESTING

Kopfler (1980) indicated several areas of concern in preparing representative concentrates for biological testing. For example, organic residues can change during storage of concentrates between preparation and biological testing or chemical analysis. In addition, humic material can bind lower molecular weight organic substances. Consequently, their recovery is necessary to ensure the integrity of the sample for such bound constituents.

Because of the large variety of chemical compounds present either naturally or as industrial contaminants in water samples, no single concentration method currently available is adequate for concentrating all organic constituents in the water sample. Consequently, in an attempt to concentrate or isolate as much of the organic matter as possible, most researchers use several methods. The combinations can become quite complex and are technically difficult to achieve. Table 4-1 summarizes most of the practical methods used to achieve concentration of organic compounds from water samples and indicates the utility of each method, along with the major advantages and disadvantages.

Several of the methods have been used to prepare concentrates for biological testing. Table 4-2 presents selected examples of these methods, along with the principal biological test used and the reference citation. Although the ability of the method to concentrate all organic matter may be limited, significant positive results were achieved in most of the studies. There is, thus, a question as to whether a single concentration method or a combination of methods can be developed to achieve concentration of all organic matter in a water sample or even if a single method is necessary. That is, it may be simpler and more economical to use several different concentration methods to concentrate and isolate organic matter for biological testing. This procedure, however, raises the possibility that a highly toxic substance may be omitted because the concentration methods used are not adequate to isolate a specific compound.

The selection of the concentration method must be based on the toxicity test to be conducted. For example, a long-term feeding study using mice could require organic material from many thousands of liters of water; a study with rats would require even more. In addition, the concentration method must be chosen on the basis of the chemical and physical properties of the organic constituents to be tested. Because of technical difficulties in the toxicological testing of highly volatile constituents, these have generally been tested as specific chemical compounds rather than as concentrates. Moderately volatile and less volatile constituents may be tested as concentrates. Concentration of volatile constituents by solvent extraction now seems appropriate for the preparation of concentrates for biological testing. Apparently, reverse osmosis, in combination with other methods as used by Kopfler et al. (1977), is currently the most useful method for preparing large quantities of nonvolatile organic concentrates. However, if thousands of liters per day must be processed, a simpler method should be developed. Thus, XAD resin or other sorption systems may be preferred. For toxicological testing, the concentration method can be tailored to isolate specific classes or individual compounds. For example, humic materials can be separated from water samples by alkaline extraction and acid precipitation.

The usefulness of toxicological tests of organic concentrates to estimate the hazards associated with water depends on the degree to which the concentrate represents the organic materials actually present in the water; that is, to estimate the total hazard, the organic concentrate should be representative of all the organic substances present in the water.

TABLE 4-1 Advantages and Disadvantages of Current Testing Methods[a]

Method	Type of Organic Compound Concentrated	Advantage	Disadvantage
Concentration Technique:			
Freeze concentration	Polar and nonpolar	Low temperature	Liter volumes, limited concentration, salt concentration
Lyophilization	Nonvolatile	Low temperature, high concentration, low contamination	1 to 100 liter volumes, salt concentration
Vacuum distillation	Nonvolatile	Ambient or near ambient temperature, high concentration, low contamination	1 to 100 liter volumes, salt concentration
Reverse osmosis	Molecular weight > 200	Ambient temperature, 100-liter volumes	Salt concentration, contamination
Ultrafiltration	Molecular weight > 1,000	Ambient temperature	Salt concentration, liter volumes
Isolation Technique:			
Solvent extraction	Nonpolar, volatile	Ambient temperature, large volumes, low salt concentration	Solvent contamination, specificity, concentrate storage
Activated carbon	Nonpolar, volatile, nonvolatile	Ambient temperature, 1,000-liter volumes	Limited recovery of adsorbed organics, concentrate storage, artifacts

Ion exchange	Polar nonpolar	Large volumes, organic recovery 70%-90%	Resin preparation and elution
XAD resin	Nonpolar to polar, volatile, nonvolatile	Convenient, ambient temperature, large volumes	Contamination, limited recovery of adsorbed organics, resin preparation
Precipitation	Humic materials, specific chemicals	Not stated	Specificity
Centrifugation	Macromolecules	Not stated	Specificity, i.e., large molecules, low volumes
Gas stripping	Volatile	Small samples	Low volumes

[a]From Jolley, 1980.

TABLE 4-2 Concentration Methods for Biological Testing

Water	Concentration Method	Reference	Biological Testing
Drinking water	Reverse osmosis	Neal, 1980	Initiation-promotion study (SENCAR mice)
Drinking water	XAD	Neal, 1980	Initiation-promotion study (SENCAR mice)
Drinking water	Reverse osmosis	Neal, 1980	Ames test
Drinking water	Reverse osmosis	Loper, 1980	BALB/3T3 fibroblasts
Drinking water	XAD	Cheh et al., 1980	Ames test
Paper plant effluents	XAD	Douglas et al., 1980	Ames test
Drinking water	Solvent extraction (combination)	Hemon et al., 1978	Cytotoxicity, promotion
Drinking water	Solvent extraction	Cabridenc and Sdika, 1979	Ames test, cytotoxicity, promotion
Wastewater effluents	Lyophilization and vacuum distillation	Cumming et al., 1979	Ames test
Wastewater effluents	Ion exchange (combination)	Baird et al., 1980	Ames test
Wastewater effluents	Ion exchange and distillation	Johnston and Verdeyen, 1981	Ames test
Drinking water	Reverse osmosis	Tardiff et al., 1978 Kopfler et al., 1977	Ames test
Drinking water	Reverse osmosis	Tabor et al., 1980 Loper et al., 1978	Ames test

RECOMMENDATIONS

Efforts should be continued to develop simple but efficient methods for concentrating all the organic materials actually present in water samples. However, meeting this goal may not be technically possible at present.

Methods to evaluate and compare different concentration techniques should be developed and applied. This effort will require increased emphasis on chemical analysis to identify much of the organic material in the concentrates. Measurements of general parameters, such as TOC, cannot be used reliably to compare concentration techniques. All concentrations techniques should be compared by determining the recovery of a series of compounds of various solubilities, selected to include a range of chemical classes, functional groups, and molecular weights, including aquatic humic substances, and these same materials should be measured before and after reaction with chlorine. Contaminants unique to the concentration process should be identified and analyzed to determine the effect of the contaminants on the toxicity test and to permit proper evaluation of that test of the concentrate itself. Possible artifact production by the concentration methods should be evaluated; for example, peroxide in ether used to elute adsorption columns may create artifacts. The stability of concentrates during storage should also be evaluated.

More specific recommendations on concentration methodologies for toxicity testing are given in Appendix A.

REFERENCES

Baird, R., J. Gute, C. Jacks, R. Jenkins, L. Neisess, B. Scheybeler, R. Van Sluis, and W. Yanko. 1980. Health effects of water reuse: A combination of toxicological and chemical methods for assessment. Pp. 925-935 in R.L. Jolley, W.A. Brungs, and R.B. Cummming, eds. Water Chlorination: Environmental Impact and Health Effects, Vol. 3. Ann Arbor Science, Ann Arbor, Mich.

Burnham, A.K., G.V. Galder, J. S. Fritz, G.A. Junk, H.G. Svec, and R. Willis. 1972. Identification and estimation of neutral organic contaminants in potable water. Anal. Chem. 44:139-142.

Cabridenc, R., and A. Sdika. 1979. Analysis of the organic micropollutants of water. Extraction of organic micropollutants from water for the purpose of performing biological assays. Paper presented at the European Symposium on Analysis of Organic Micropollutants of Water, Berlin, Dec. 11-13, 1979. EPA translation No. TR80-0252. U.S. Environmental Protection Agency, Washington, D.C.

Cheh, A.M., J. Skochdopole, P. Koski, and L. Cole. 1980. Nonvolatile mutagens in drinking water: Production by chlorination and destruction by sulfite. Science 207:90-92.

Cumming, R.B., L.R. Lewis, R.L. Jolley, and C.I. Mashni. 1979. Mutagenic activity of nonvolatile organics derived from treated and untreated wastewater effluents. Pp. 246-252 in A.D. Venosa, ed. Progress in Wastewater Disinfection Technology. Proceedings of the National Symposium, Sept. 18-20, 1978. EPA-600/9-79-018. U.S. Environmental Protection Agency, Cincinnati, Ohio.

Deinzer, M., R. Melton, and D. Mitchell. 1975. Trace organic contaminants in drinking water: Their concentration by reverse osmosis. Water Res. 9:799-805.

Douglas, G.R., E.R. Nestman, J.L. Betts, J.C. Mueller, E.G.-H. Lee, H.F. Stich, R.H.C. San, R.J.P. Brouzes, A.L. Chmelauskas, H.D. Paavila, and C.C. Walden. 1980. Mutagenic activity in pulp mill effluents. Pp. 865-880 in R.L. Jolley, W.A. Brungs, and R.B. Cumming, eds. Water Chlorination: Environmental Impact and Health Effects, Vol. 3. Ann Arbor Science, Ann Arbor, Mich.

Fan, T.Y., R. Ross, D.H. Fine, L.H. Keith, and A.W. Garrison. 1978. Isolation and identification of some thermal energy analyzer (TEA) responsive substances in drinking water. Environ. Sci. Technol. 12:692-695.

Fang, H.H.P., and E.S.K. Chian. 1976. Reverse osmosis separation of polar organic compounds in aqueous solution. Environ. Sci. Technol. 10:364-369.

Gustafson, R.L., and J. Paleos. 1971. Interactions responsible for the selective adsorption of organics on organic surfaces. Pp. 213-237 in S.J. Faust and J.V. Hunter, eds. Organic Compounds in Aquatic Environments. Marcel Dekker, Inc., New York.

Hemon, D., P. Lazar, R. Cabridenc, I. Chouroulinkov, A. Sdika, B. Festy, and C. Gerin-Roze. 1978. Organic micropollution of waters intended for human consumption. Rev. Epid. Sante Pub. 26:441-450.

Hindin, E., P.J. Bennett, and S.S. Narayanan. 1969. Organic compounds removed by reverse osmosis. Water Sewage Works 116:466-470.

Johnston, J.B., and M.K. Verdeyen. 1981. Recovery of mutagens from water by the Parfait/distillation method. Pp. 171-190 in W.J. Cooper, ed. Chemistry in Water Reuse. Vol. 2. Ann Arbor Science, Ann Arbor, Mich.

Jolley, R.L. 1980. Concentration of organic compounds. Workshop on Protocol Development: Criteria and Standards for Potable Reuse and Feasible Alteratives, Airlie House, Va., July 29-31, 1980. U.S. Environmental Protection Agency, Washington, D.C.

Junk, G.A., J.J. Richard, M.D. Grieser, D. Witiak, J.L. Witiak, M.D. Arguello, R. Vick, H.J. Svec, J.S. Fritz, and G.V. Calder. 1974. Use of macroreticular resins in the analysis of water for trace organic contaminants. J. Chromatog. 99:745-762.

Kawabata, N., and T. Morigaki. 1980. Removal and recovery of organic pollutants from the aquatic environment. 2. Removal and recovery of dodecylbenzenesulfonate from aqueous solution by cross-linked poly(N-benzyl-4-vinylpyridinium halide). Environ. Sci. Technol. 14:1089-1093.

Klein, E., J. Eichelberger, C. Eyer, and J. Smith. 1975. Evaluation of semi permeable membranes for determination of organic contaminants in drinking water. Water Res. 9:807-811.

Kopfler, F.C., W.E. Coleman, R.G. Melton, R.G. Tardiff, S.C. Lynch, and J.K. Smith. 1977. Extraction and identification of organic micropollutants: Reverse osmosis method. Ann. N.Y. Acad. Sci. 298:20-30.

Kopfler, F.C. 1980. Alternative strategies and methods for concentrating chemicals from water. Presented at the Second

Symposium on Application of Short-Term Bioassays in the Fractionation and Analysis of Complex Mixtures. Environmental Protection Agency, Washington, D.C.

Lawrence, J., and H.M. Tosine. 1976. Adsorption of polychlorinated biphenyls from aqueous solutions and sewage. Environ. Sci. Technol. 10:381-383.

Loper, J.C. 1980. Overview of the use of short-term biological tests in the assessment of the health effects of water chlorination. Pp. 937-945 in R.L. Jolley, W.A. Brungs, and R.B. Cumming, eds. Water Chlorination: Environmental Impact and Health Effects, Vol. 3. Ann Arbor Science, Ann Arbor, Mich.

Loper, J.C., D.R. Lang, R.S. Schoeny, B.B. Richmond, P.M. Gallagher, and C.C. Smith. 1978. Residue organic mixtures from drinking water show in vitro mutagenic and transforming activity. J. Toxicol. Environ. Health 4:919-938.

Neal, R.A. 1980. Known and projected toxicology of chlorination by-products. Pp. 1007-1017 in R.L. Jolley, W.A. Brungs, and R.B. Cumming, eds. Water Chlorination: Environmental Impact and Health Effects, Vol. 3. Ann Arbor Science, Ann Arbor, Mich.

Smith, C.C. 1978. Strategy for collection of drinking water concentrates. Pp. 227-244 in M.D. Waters, S. Nesnow, J.L. Huisingh, S.S. Sandbu, and L. Claxton, eds. Application of Short-Term Bioassays in the Fractionation and Analysis of Complex Environmental Mixtures. Plenum Press, New York.

Tabor, M.W., J.C. Loper, and K. Barone. 1980. Analytical procedures for fractionating nonvolatile mutagenic components from drinking water concentrates. Pp. 899-912 in R.L. Jolley, W.A. Brungs, and R.B. Cumming, eds. Water Chlorination: Environmental Impact and Health Effects, Vol. 3. Ann Arbor Science, Ann Arbor, Mich.

Tardiff, R.G., G.P. Carlson, and V. Simmon. 1978. Halogenated organics in tap water: A toxicological evaluation. Pp. 195-209 in R.L. Jolley, ed. Water Chlorination: Environmental Impact and Health Effects, Vol. 1. Ann Arbor Science, Ann Arbor, Mich.

Thurman, E.M., R.L. Malcolm, and G.R. Aiken. 1978. Prediction of capacity factors for aqueous organic solutes adsorbed on a porous acrylic resin. Anal. Chem. 50:775-779.

5
Health Effects Testing

The adequacy of a water treatment facility rests on its ability to produce water that, according to chemical and toxicological procedures, is at least as free of adverse health effects or risks to humans as are conventional water sources. Currently, potable supplies must meet a variety of primary and secondary standards and also originate from a relatively nonpolluted source, either surface or ground. Although present drinking water standards (Environmental Protection Agency, 1979a,b) undoubtedly apply as a minimum to a reused water, a problem arises when evaluating reused water because it does not originate from an "acceptable" source as defined by current regulations. Clearly, prudence demands that if reused water is to be considered for human consumption it must be shown beyond a reasonable doubt that it poses no greater risk than water from conventional, less contaminated sources.

This chapter reviews the information on health effects testing of effluents from existing water reuse facilities and epidemiological studies. Then a scheme is presented for identifying potential adverse health effects of reused water using conventional water supplies for comparisons.

The Safe Drinking Water Act of 1974 was passed in part because of an increasing awareness that drinking water in the United States contained a wide variety of potentially harmful substances. Publications from the National Academy of Sciences (1977a, 1980a, 1982) elucidated health effects from exposure to a large number of individual inorganic, organic, and microbiological contaminants. By contrast, simultaneous exposure to complex mixtures of these and other chemicals continues to be more difficult to assess.

In classic toxicological procedures, the effects of a relatively pure chemical substance are measured in groups of living organisms against a control group, which is subjected to the same procedure but without exposure to the test chemical. If wastewater or treated wastewater were to contain only one chemical, toxicity testing would be straightforward; but drinking water, conventional or reused, contains complex mixtures of chemicals that greatly complicate health effects testing and risk assessment. People who drink water are consequently exposed to a wide variety of chemicals. Even if they could be identified, testing all the chemicals present in such mixtures is a practical impossibility. Therefore, for practical reasons it becomes necessary to conduct toxicological tests on complex mixtures.

Although it may be possible to make judgments about the absolute safety of an individual chemical, it is doubtful that this can reasonably be done for exposures to complex mixtures. If it is assumed that the risks from conventionally treated water are generally acceptable, then it may be possible to make comparisons between two supplies based on the results of comparative chemical, microbiological, and toxicity testing. Consequently, the Panel on Quality Criteria for Water Reuse believes the issue of comparative risk to be central to any meaningful decision regarding acceptability of reused supplies.

Options for water reuse often include wastewater as one of several potential sources for potable water. The final selection of source is based on cost, if no excess adverse health risks are known to be associated with the less expensive source. For such situations, it may be practical to compare adverse health effect differences between potable water prepared from conventional sources and water prepared from treatment of wastewater effluent blended with source water. Of course, the conventional source used for such comparison may be subject to water quality degradation and may contain health risks at any time, even if the source is presumed to be "safe." A comparative approach, therefore, may not produce an easily interpreted outcome if toxicity tests of the conventional source produce positive results. An additional problem is presented by the need to make public investment decisions for future water supplies and treatment systems, based on present quality estimates. Therefore, comparative testing must be reasonably frequent and the results closely scrutinized by well-qualified professionals.

Risk evaluation rests on a series of toxicological procedures, which rely, in large measure, on responses in whole animals. Traditionally, only single chemicals are evaluated in such tests. However, reused water and conventionally treated waters contain an extensive mixture of complex and largely unidentified chemicals, and, therefore, the toxicological testing recommended here involves evaluating the effects of exposure to mixtures in several systems, including whole animals. The panel realizes that this approach represents a departure from traditional toxicological procedures, but such test conditions more closely represent actual human exposures. Furthemore, the panel recognizes the generality of this need with respect to the testing of health effects of samples from any environmental medium.

REVIEW OF HEALTH EFFECTS FROM AND TESTING OF REUSED WATER

Evaluation of Existing Reuse Facilities

Hattingh (1978) and Van Rensburg et al. (1977) reported on the operation and toxicity testing of the (unconcentrated) reclaimed water from the Stander plant in South Africa. Their measurements and tests covered an operational period of 2 years and included an extensive battery of microbiological, chemical, and toxicological tests.

Microbiological tests included measurements of total bacterial plate count, total coliform count, fecal coliform count, fecal streptococci, Clostridium perfringens, Staphylococcus aureus, Pseudomonas aeruginosa, and counts of viruses and parasitic ova. Chemical tests included measurements of heavy metals and all of the usual chemical parameters that are measured as indicators of water quality, in addition to determinations of organochloro and organophosphorus compounds, polynuclear aromatic hydrocarbons (PAH's), and organic substances concentrated on XAD resin. Toxicity tests were performed on rats exposed subcutaneously and orally for 30 months. Different waters were evaluated, including humus tank effluent (settled biological filter effluent), reclaimed water, tap water (potable), and distilled water. Aliquots were fed to rats as their sole source of liquid. In addition, spent activated carbon and virgin activated carbon were mixed into the feed of other test groups. Furthermore, the rats were bred and observed for alterations in reproductive performance.

The results of these tests indicated that the treatment processes used in this facility were very effective in reducing potential microbiological and chemical toxicity. Depending on the degree of treatment, microbiological and viral activity was reduced from 1 to 7 logs and was comparable in quality to that of other potable waters in the Pretoria area. The quantities of heavy metals and other chemical indices were reduced to within the numerical levels of prevailing drinking water standards. Concentrations of the classes of organic compounds listed above were reduced by 84%.

The only water fraction that had a harmful effect on the rats was the humus tank effluent; rats ingesting this fraction had an increased incidence of gastrointestinal infection due to viruses. Some rats that received activated carbon in their feed and drank tap water developed adrenal medullary hyperplasia. Breeding performance and estrous cycle activity were normal in all groups. Hattingh (1978) concluded, "The results obtained indicate that the reclaimed water was of high quality and conformed to all known potable water quality standards. Bioassay results indicate that reclaimed water had no deleterious effects on rats or fish."

The microbiological and chemical quality of reclaimed water from Water Factory 21 has been critically evaluated (McCarty et al., 1980a,b), but, to date, there has been no toxicity testing. All inorganic and organic chemicals monitored in the reclaimed water were below existing standards or guidelines for at least 98% of the time during a 15-month period. In addition, various aromatic hydrocarbons, synthetic chlorinated compounds, chlorination products, natural products, phthalate esters, and miscellaneous compounds were monitored. Viruses were found in the effluent water only twice; in each instance their presence was attributed to the operation of the activated carbon towers in an upflow mode. None was detected when the towers were operated in a downflow mode. Results of other standard microbiological tests (for total coliforms and fecal coliforms) indicated that the treatment process was successful in producing water that met existing microbiological criteria.

The Experimental Estuary Water Treatment Plant (EEWTP) at Blue Plains in Washington, D.C., has recently become operational. Plans for this facility include a number of chemical, microbiological, and toxicological tests. Chemical tests include measurements of heavy metals, standard water quality indices, and organic priority pollutants (Army Corps of Engineers, 1981). Microbiological tests will include measurements of fecal coliforms, total coliforms, parasites, endotoxins, and viruses. The toxicity tests are limited to two short-term *in vitro* assays: Ames/*Salmonella* and a mammalian cell transformation assay with the C3H/10T1/2 mouse embryo fibroblast system. These *in vitro* tests will be performed on water samples concentrated on, and eluted from, XAD macroreticular resins (Army Corps of Engineers, 1981). Gruener (1978, 1979) has already performed extensive toxicological tests on concentrated effluent from the Blue Plains plant. His results are reviewed later in this chapter.

Results of Epidemiological Studies

The effects on human health (e.g., cancer and cardiovascular disease) that may be associated even with conventional water ingestion are extremely difficult to measure. Studies in animals in which high doses of organic compounds were used have suggested that some of those found at trace levels in drinking water may have carcinogenic potential (National Academy of Sciences, 1977a, 1980b). The only way to assess the effects of these organic compounds in human populations is to conduct epidemiological studies. Such studies are difficult, expensive, and subject to a high degree of uncertainty. A critical perspective on the utility of epidemiology to elucidate the potential effects from the kinds of studies cited below is offered in a recent publication from Doll and Peto (1981). In their discussion of the limitations of epidemiology, they stated:

> Trustworthy epidemiological evidence, it should be noted, always requires the demonstration that a relationship holds for individuals (or perhaps small groups) within a large population as well as between large population groups. Correlation between the incidence of cancer in whole towns or whole countries and, for example, the consumption of particular items of food can, at the most, provide hypotheses for investigation by other means. Attempts to separate the role of causative and confounding factors by the statistical techniques of multiple regression analysis have been made often, but evidence obtained in this way is, at best, of only marginal value.

With respect to human population studies in which positive correlations between the amounts of certain contaminants and mortality from certain cancers have been reported, Doll and Peto concluded:

> . . . the interpretation of the correlation studies is at present open to question. Similar studies have been carried out over the years in many other fields, but have rarely been regarded as constituting anything more than

hypotheses to be tested by more specific work, because of the difficulty of obtaining truly relevant data (relating to long-past exposure of the actual individual concerned) and of eliminating the effects of concomitant variation. Analyses that took account of other important variables and were consistent from one region to another in pointing to specific effects on one or other specific type of cancer would carry some weight, but most of the analyses have not met these criteria.

The Safe Drinking Water Committee (National Academy of Sciences, 1980a) reviewed 12 such studies (most of which were the ecological type) and concluded that they:

> ... failed either to support or to refute the results of positive animal bioassays suggesting that certain trihalomethanes (THM's), e.g., chloroform, may cause cancer in humans. Any association between THM's and bladder cancer "was small and had a large margin of error"—not only because of statistical variances but also, much more importantly, because of the very nature of the studies. All of the epidemiological studies were handicapped by the extreme difficulty of identifying a very small effect in a population. The methodological complexities inherent in epidemiological studies of human populations exposed to multiple contaminants at low concentrations (ppb) in drinking water make it virtually impossible to establish a causal link between THM's and an increase in cancer of the bladder or of any other site. Small differences in cigarette consumption between two population groups could account for the observed associations. Any causal relationships between THM's and bladder cancer are weakened by imprecise exposure data. In most of the studies, THM concentrations in different water sources were only inferred, rather than actually measured. In addition there are difficulties in controlling for a multitude of factors that are known to affect cancer incidence: cigarette smoking, diet, occupation, use of alcohol and drugs, socioeconomic status, ethnicity, and nonaqueous sources of THM's.

Slightly different conclusions were reached by Crump and Guess (1980) in a subsequent review prepared for the Council on Environmental Quality. In this report, the authors reviewed more case-control studies than did the Safe Drinking Water Committee. These studies represent a degree of refinement over earlier ecological studies, but are still limited in several major ways, i.e., indirect measure of water quality (chlorinated versus unchlorinated, or surface versus ground), and use of death certificate data (no information on dietary or smoking habits, occupational history). Crump and Guess concluded:

1. The recently completed case control studies have strengthened the evidence for an association between rectal, colon and bladder cancer and drinking water quality provided by the earlier epidemiological studies reviewed by the National Academy of Sciences committee. While the epidemiological studies completed to date are not sufficient to establish a causal relationship between chlorinated organic contaminants in drinking water and cancer, they do contain evidence which supports such a relationship for rectal cancer and, to a lesser extent, for bladder and colon cancer.

2. Putative increases in cancer risks associated with organic contaminants in drinking water appear to lie near the low limit of what can be reliably detected by epidemiological methods.

3. No clear trend of increasing cancer risk with increasing exposure to organic contaminants in drinking water has been demonstrated by the studies conducted to date although evidence suggestive of such trends has been obtained for rectal cancer in one study and for colon cancer in another study.

4. Concentrates of chlorinated nonvolatile organic compounds in drinking water have been found to be mutagenic in mammalian cells and to be capable of transforming human cells into cells which exhibit some biochemical properties associated with tumor cells. These results support the hypothesis that chlorinated nonvolatile organic compounds in drinking water may be carcinogenic in humans. Most of the nonvolatile organic content of drinking water has not yet been identified.

To date, few epidemiological studies have been conducted on persons drinking reused water. In one study, Grabow and Isaacson (1978) investigated the microbiological quality and waterborne diseases associated with reclaimed water from a treatment plant at Windhoek, Namibia. They studied the occurrence of communicable diseases from organisms such as <u>Salmonella</u>, <u>Shigella</u>, enteropathic <u>Escherichia coli</u>, <u>Vibrio</u>, enteroviruses, and <u>Schistosoma</u>, as well as diseases such as viral hepatitis, meningitis encephalitis, and nonbacterial enteritis. The epidemiological studies did not reveal any differences in disease rates among people consuming water from either reused or conventional supplies.

More recently, Nellor <u>et al</u>. (1981) and Frerichs <u>et al</u>. (1981) reported preliminary results of an epidemiological study of populations in the Montebello Forebay region of Los Angeles, which received some reused water as part of the drinking water supply. These people received reused water as a result of ongoing programs to recharge groundwater supplies and also to provide a freshwater barrier against the intrusion of seawater. Populations in four geographic regions were compared with controls (no exposure); two groups with low exposure received water containing less than 5% reclaimed water for at least 1 year prior to 1970 and two groups with high exposure received water containing more than 5% reclaimed water for at least 1 year prior to 1970.

Twenty possible adverse health effects were examined: death from cancer at various sites and other causes, birth and fertility data, infant mortality, and morbidity including potential waterborne diseases. Of the 20 indices examined, 5 showed statistically significant differences among the four geographic areas. However, none of the differences was consistent with the biological hypothesis that ingestion of reused water would cause adverse health effects. Only for rectal cancer was there an increased incidence of death in the high-exposure area as compared with the low-exposure area. This difference was of a low order and could have been attributable to chance ($\underline{P}$ = 0.08). The authors concluded that, as of 1969-1971, there were no grossly apparent adverse health effects associated with the consumption of reused water.

The same investigators are continuing their study in three modes: analyzing the defined populations for the years 1972-1978 to determine if continuing exposure to reused water may result in a change in the 20 measured adverse health effects; conducting an intensive health survey of 1,250 adult women residing in a reused water area and comparing results with a study of 1,250 adult women in a control area to measure more sensitive health indices that cannot be gleaned from vital statistics data; and refining the mathematical model. The model attempts to relate such factors as low-dose exposure; long latency periods; significance of positive as well as negative findings; and the size, composition, and mobility of the study populations. It is anticipated that this study will be completed sometime in 1982.

Although it would be useful either to prove or to disprove the alleged association between contaminated drinking water and cancer, none of the studies mentioned above has sufficient statistical sensitivity to meet this end. Unless epidemiological methodology is improved, it is doubtful whether it can be used to evaluate the potential carcinogenic risk of drinking reused water. On the other hand, it does seem reasonable to monitor for various waterborne infectious diseases, because such acute effects are more easily detected and capable of being associated with their causative agents and sources of exposure.

To date, the <u>limited</u> toxicity tests performed on reused water and epidemiological studies of exposed populations have not shown that consumption of reused water represents any greater or lesser risk than does consumption of water from other conventional sources.

Results from the Toxicity Testing of Mixtures

<u>In Vitro</u> Studies

The general concept of testing and evaluating the potential health effects of mixtures is difficult and complex. It has long been recognized that humans are both acutely and chronically exposed to a large number of chemicals, but scientific attempts to measure potential effects have been meager. Logistics and economics, as well as the toxicological community's perception of a proper evaluation of toxicity, have prevented a systematic study of the health consequences from exposure to complex mixtures. In general, there is little information identifying the chemicals present in environmental mixtures. The subject of testing mixtures has been reviewed (National Academy of Sciences, 1980 a,b), but the data have primarily dealt with mixtures of only two or three components.

There are several studies of the mutagenic potential of unconcentrated drinking water supplies. Pelon et al. (1980) used both raw and finished drinking water collected between 1974 and 1976 from the Mississippi River in southeastern Louisiana. They measured the transforming ability of these samples on the mouse clonal cell line R846-DP8. Transformation was seen in 7 of 118 (6%) Mississippi River water samples, 7 of 70 (10%) raw water samples, and 5 of 115 (4%) finished water samples. The authors concluded that limitations in methodology or their criteria for transformation may have precluded a higher proportion of positive results.

Grabow et al. (1981) measured the mutagenic activity of unconcentrated raw and treated water from the Vaal River in South Africa. Their results indicated that there was more mutagenic activity in the treated water than in the raw river water; most mutagenicity was expressed without metabolic activation. The authors attributed the increased mutagenicity to the volatile and nonvolatile organic compounds formed during water treatment. Furthermore, they suggested that, under certain circumstances, total organic carbon (TOC) may serve as a useful indicator of potential mutagenic precursors. Correlations of this kind should be investigated further, because they may be useful for routine monitoring. (See Chapter 4.)

Loper (1980) reviewed the subject of testing concentrated water samples and the deficiencies of using unconcentrated samples. He discussed the variability of test results, depending on which concentration method is used and the lack of information about specific compounds in the drinking water residues. He concluded:

> The future identification of even a few of the actual mutagens involved will permit a systematic analysis for answers to such questions as the contribution of point sources to the origin of the mutagens and the effects of alternate disinfection treatments, seasonal effects, the optimal concentration and identification procedures for monitoring purposes, possible hazards to health, and procedures for avoidance or elimination of the mutagens.

Recently, Loper (personal communication, 1981) tentatively identified a particular compound responsible for some of the mutagenicity of a drinking water concentrate. Using high-pressure liquid chromatography (HPLC), he isolated the compound 1,3-dichloro-2-propenyl-2-chloroethyl ether from a complex mixture. Although the source of this compound is not known, its identification may, for the first time, permit an attempt to locate point sources or treatment techniques responsible for its presence.

Nellor et al. (1981) attempted to fractionate and identify mutagenic activity from concentrated well water in the Montebello Forebay region of Los Angeles County. These wells, recharged over the past 18 years with secondary and tertiary effluent, supply a substantial quantity of groundwater for distribution by domestic suppliers. In addition to testing concentrates for mutagenicity using Salmonella typhimurium strains TA98 and TA100, the authors also separated up to 60 fractions and measured their mutagenic activity. Nellor et al. suggest that most of the mutagenicity was due to the additive effects of rather low molecular weight compounds. They identified N-nitrosomorpholine, N-nitrosopiperidine, N-nitroso-N-methylethaneamine, and benzo[a]pyrene, but not in concentrations that, individually, would account for the mutagenicity observed. Tentatively identified were various plasticizers such as phthalic acid, diethyl phthalate, dibutyl phthalate, tributyl phosphoric acid, and isopropyl myristic acid and various petroleum byproducts such as alkyl benzenes, alkyl naphthalenes, alkyl phenols, alkyl phenanthrenes, aldehydes, glycol ethers, and aliphatic amines. Several unknown compounds, at least two of which were brominated, were observed in apparently high concentrations in the mutagenic fractions of well water samples. This work is being continued in an attempt to characterize these compounds more fully.

Tabor et al. (1980) reported that disinfection (ozonation or chlorination) frequently increased the mutagenic activity of drinking water concentrates when measured by Salmonella tester strains TA1535, TA1538, TA98, and TA100. Most of the mutagens were direct acting, and the addition of a microsomal-activating system (S9) decreased mutagenic activity. Both base-pair-substitution and frameshift mutagens were found in wastewater, the former more frequently. Following fractionation with HPLC, the mutagenic activity of the sample often increased, which may be indicative of interactions between components in these complex mixtures.

Similar observations were made by de Greef et al. (1980), who noted that disinfection by either chlorine or chlorine dioxide increased the mutagenic activity of XAD water concentrates as measured by the Salmonella/microsome assay. Addition of S9 significantly increased mutagenic activity of raw water concentrates, but had little or no effect on treated water concentrates. The authors postulated that chlorine dioxide oxidation "mimicked" the action of S9 mixed-function oxidases.

McCarty et al. (1980a) also reported an increase in mutagenic activity after disinfection with chlorine. They studied wastewater concentrates (on XAD resins) from the Palo Alto Reclamation Plant on Salmonella strains TA98, TA100, TA1535, TA1537, and TA1538 with and without S9 activation. Mutagenic activity was usually found in the influent water, especially to strains TA98 and TA1535 with S9 activation. The only treatment process that removed this activity was adsorption using activated carbon. Chlorination of the treated effluent resulted in an increase in mutagenic activity, especially to strain TA1535; chlorine dioxide did not appear to increase mutagenic activity.

Most of the initial screening of drinking water concentrates has been done in the Salmonella/microsome assay. Although this test has relatively good correlation with carcinogenicity, the use of a battery of short-term tests is believed to enhance the predictability of potential carcinogenicity (McCann et al., 1975). To this end, Lang et al. (1980) investigated the transformation of BALB/3T3 cells by drinking water concentrates from five U.S. cities. The authors reported various degrees of positive results, i.e., transformation of 3T3 cells, from all five cities. Virtually all of the positive transformation data came from samples concentrated by reverse osmosis (RO), rather than by XAD resins.

Lang et al. confirmed that the transformed cells were malignant by injecting them into athymic nude mice and evaluating their potential for tumor formation. All transformed cells were able to induce tumors within an average latency period of 28 days. Although the interpretation of such experiments is difficult due to concentration effects (artifacts) and overt cellular toxicity, these studies do extend the ability to assess the potential health consequences from drinking water concentrates.

In mutagenicity tests using water concentrates containing up to 1,000 times the expected exposure of humans to TOC, Gruener (1979) reported no significant effect on Salmonella strains TA1535, TA1537, TA98, and TA100 and a clearly mutagenic S9-dependent effect on V79 hamster cells. In addition, he demonstrated that human lung fibroblasts (WI38) were sensitive to the effects of water concentrates. Ranges in protein levels were used to demonstrate toxicity in this assay. The presence of the S9-activation system increased the toxic effect of the water concentrates.

In another short-term assay, Gruener and Lockwood (1980) found freeze-dried concentrates to be mutagenic in a mammalian tissue culture assay using Chinese hamster embryonic lung cells (V79). Increased rates of mutagenesis occurred only after activation with a liver microsomal system. A preliminary test with a concentrate containing only 140 mg/liter of TOC showed no activity.

Aside from the question of the significance of the various in vitro tests, one important caution should be taken concerning the methodology of preparing samples of chlorinated water supplies. Such samples are often dechlorinated with reducing agents prior to chemical analysis and/or toxicity testing. Recently, Cheh et al. (1980) showed that dechlorination (such as by sulfate) can reduce mutagenicity in a model water treatment system. Thus, in the development of protocols for testing treated wastewater supplies, it is important to ensure that chlorinated compounds or other mutagens or toxicants actually present are not inadvertently modified in sample preparation.

In Vivo Studies

Tardiff and Deinzer (1973) reported some preliminary in vivo acute toxicity studies in which they determined LD_{50} values in mice from RO concentrates. The LD_{50} values varied from 32 mg/kg to 290 mg/kg, depending on the final extraction procedure used. This range makes comparisons and data interpretation difficult.

A battery of in vivo tests was used by Robinson et al. (in press) to evaluate RO and XAD concentrates from the drinking water of five cities. They measured the initiating activity of the concentrates following subcutaneous injection into SENCAR (sensitive-to-carcinoma) mice. The residues were injected six times over a 2-week period, followed by topical application of phorbol myristate acetate (PMA) three times weekly for 20 weeks. The positive control was dimethyl-benzanthracene (DMBA) plus the PMA. After 50 weeks, there were significantly more papillomas per mouse in the animals treated with the RO and XAD concentrate samples from several cities. Tumor-promoting potential was tested by initiating with topical DMBA, followed by 20 weeks of promoting with the RO and XAD concentrates; the positive control was DMBA followed by PMA. After 38 weeks, there were no tumors in the test groups and a total of 319 papillomas in the 20 mice used as positive controls. Neither the RO nor the XAD concentrates had any effect after 38 weeks when tested alone (applied to the skin) to measure their potential as complete carcinogens.

Another short-term in vivo test proposed to measure the potential carcinogenicity of drinking water concentrates is the rat liver foci assay. This test, as described by Ford and Pereira (1980), uses nitrosodiethylamine (NDEA) as the initiator, with phenobarbital and partial hepatectomy as promotors. Initiation was determined as the induction of hepatic foci of γ-glutamyl transpeptidase (GGTase) activity. GGTase-positive foci have been associated with a high percentage of phenotypically altered foci. For the purpose of this assay, it is assumed that such foci progress to nodules and, ultimately, to cancer and are therefore molecular markers for the initiation of cancer. However, the monoclonal nature of cells needs to be demonstrated, as well as the irreversible nature of such foci after administration of the test substance is discontinued. This test showed good dose dependency and sensitivity down to 0.003 mmol/kg of NDEA. This assay is currently being evaluated for use in a matrix of short-term tests for evaluating the potential carcinogenic hazard of drinking water concentrates (R. Bull, personal communication, 1981).

Gruener (1979) performed subchronic toxicity evaluations of water concentrates containing up to 1,000 times the expected human exposure to TOC. He exposed five groups of CFl mice in the following manner: exposure for 14 days (50 males, 100 females); exposure for 90 days (200 males, 200 females); exposure during gestation, lactation, and another 90 days (100 males, 100 females); exposure during gestation, lactation, and another 150 days (50 males, 50 females); and exposure for 90 days followed by another 90 days on regular diet (50 males). The results of the many different tests performed during the various studies showed no significant differences in hematological values, motor activity, body weight, or mixed-function oxidase activity. Also, no differences were noted in any of the reproduction tests.

Studies of the possible harmful effects of water constituents on mammalian reproduction and embryonic growth and development have been few and inadequate. To date, only five such studies have been performed, three in which the effects of a municipal tap water were compared with those of purified water (Chernoff et al., 1979; McKinney et al., 1976; Staples et al., 1979) and two others in which water concentrates were tested (Gruener, 1978; Kavlock et al., 1979).

In the study by McKinney et al., an increased frequency of prenatal mortality and certain malformations was attributed to consumption of tap water during 1 month of the study period. This result prompted later repeat studies of the same water source by Staples et al. (1979) and Chernoff et al. (1979), who failed to confirm that the water had any untoward effects on reproduction and fetal health. Gruener (1978) conducted male and female reproduction tests (dominant lethal), but no teratological studies, with several dilutions of a water concentrate incorporated into the diet. He found inconsistent effects, and no dose-response relationships, on litter size or offspring weight. Amounts of concentrate administered were calculated to be 100 to 1,000 times the expected human exposure levels. In a teratology experiment, Kavlock et al. (1979) administered concentrates from drinking waters of five U.S. cities and a synthetic mixture of volatile substances by gavage to mice at 300 to 3,000 times the expected human exposure level. The results were negative.

None of the studies cited above has yielded definitive information on the potential effects of substances in water on prenatal development. Only one of the studies used concentrates to test for teratological effects, and this did not extend the test dosages to the end point of frank toxicity. This lack was probably due to the limited supplies of concentrates. Another shortcoming of the five studies was that they all used only one species (mice) and all but one used the same outbred stock. Conventional protocols recommend that at least two species be used for teratological testing.

The ultimate evaluation of the potential adverse health effects from reused water must come from chronic bioassays in whole animals. Lifetime feeding studies to detect carcinogenicity using a maximum tolerated dose (MTD) with even a single chemical are often difficult to interpret with regard to anticipated risks in humans. These problems can be greatly magnified when the test material is a complex, undefined mixture.

The microbiological in vitro studies using drinking water concentrates cited here have all demonstrated a positive mutagenic effect. Although a battery of short-term tests has demonstrated good correlation with carcinogenicity, this result has been validated only for individual chemicals. Because of the complex nature of water concentrates, the assumption that positive results in short-term tests are predictive of carcinogenicity may or may not be valid. Artifact formation in the concentrate or interactions such as synergism may produce effects in in vitro systems that would not be seen in whole animal studies.

Dose-response models used to extrapolate the carcinogenic effects of high doses to low doses in animals are all based on exposure to a single compound. The use of such models is controversial, and applying them to mixtures increases the degree of complexity. However, they still represent the best available tools for estimating chronic risks to human health (National Academy of Sciences, 1977a, 1980b).

Toxicological Assessment

The panel considers it a practical impossibility to evaluate thoroughly and compare the different health effects of reused water with those of conventional water, based on analysis of individual compounds alone. Such data cannot predict health effects responses to actual combinations of chemicals. Moreover, there are no data on the health effects of most chemicals known to be present in water. The number of compounds already identified in drinking water supplies, although large (> 1,000), has been estimated to represent only 10% to 15% of the TOC known to be present. Therefore, it is not possible to prepare complete, synthetic mixtures for use in whole animal studies, and the only means available to obtain genuinely representative mixtures is through the preparation of concentrates. Results from the testing of mixtures in animals are difficult to use for risk evaluation because the mathematical quantitative risk extrapolation models are designed to assess exposure only to single chemicals. The following factors assume experimental importance because of the complex and essentially unverifiable nature of the mixtures involved:

- The changing consistency of samples may effect the reproducibility of results.
- Additive, synergistic, or antagonistic effects of mixture components could vary with individual samples and change with time.
- Concentration procedures may influence the chemical or physical composition of samples.
- Chemical and physical stability of concentrates may vary over time.
- Mechanics of sample preparation for administration to animals may also exert an influence.

Toxicological evaluations are normally conducted on materials of known (adequately characterized) composition and stability. The test substance should be the same as that to which the public will be exposed. If the source of the water, the treatment process, or both change, the end product must be subjected to appropriate toxicological testing to determine if toxicity differs from that of the original test material.

It is essential that the risks associated with the treated wastewater be compared with those of the conventional supplies. To achieve this objective, the panel recommends a tiered testing approach, which is described later in this chapter. Such an approach will provide a rapid indication of relative toxicity and an opportunity for comparative risk assessment in a cost-effective manner. Because of the relative insensitivity of the classical models for toxicological assessment, the water must first be concentrated before undertaking the tiered test. During the evaluation of a pilot plant effluent, concentrates must be prepared from samples obtained on a continuing basis so that they will contain representative quantities of constituents to which humans will be typically exposed. If preliminary *in vitro* and *in vivo* studies indicate adverse toxicity and additional testing is indicated, an adequate sample of the same water must be available for that testing.

Sample Collection and Preparation

If the treatment process is varied after the initial sample was prepared, a concentrate prepared at a later date may not have the same chemical characteristics as the original. This variance underscores the necessity of establishing some means by which such complex mixtures may be compared grossly, if not in detail. Current methodology would allow at least a partial "chemical fingerprint" to be established using advanced gas chromatography-mass spectrometry (GC-MS) techniques. There may be site-specific chemical tracers identified with this technique. These "sentinel chemicals" would constitute a measure of sample variance. Although the various GC-MS peaks need not be identified initially, their presence could provide a qualitative index for comparison with subsequent samples.

If GC-MS fingerprints of individual compounds in the concentrates are established for reused and conventional water supplies, it may be possible to compare the relative toxicities and potential risks of the supplies on an individual compound basis recognizing that the available lists of compounds are likely to be limited. The Safe Drinking Water Committee (National Academy of Sciences, 1977a, 1980b) and, more recently, the U.S. Department of Health and Human Services (DHHS) (Helms et al., 1981) have published lists of potentially harmful chemicals found in drinking water. The DHHS publication classifies 767 organic chemicals as recognized carcinogens, suspected carcinogens, tumor promoters, cocarcinogens, or mutagens. In a comparative analysis, water that contains twice as many recognized carcinogens as another might be judged more hazardous, assuming equal concentrations of the compounds present in both supplies and that identified potential effects are additive. Comparisons based on the presence of tumor promoters or mutagens would not be as strong, because less is known about the consequences from exposure to such chemicals. Thus, this approach could be used only within the context of another, more comprehensive analysis. Theoretical models to test for the joint actions of two or more agents have been reviewed by the Safe Drinking Water Committee (National Academy of Sciences, 1980a), which concluded:

> The models of joint toxic action could be of benefit in the risk assessment of low exposures; however, none has been adequately studied for this purpose. Their theoretical and practical implications need to be studied further before their utility can be assessed. High dose to low dose extrapolations for individual agents is an unresolved problem filled with many unknowns, and extrapolation of the actions of joint agents contains an additional major source of uncertainty.

Given the above uncertainties with respect to predicting interactions from information about single agents, it would be desirable to have in vitro and in vivo systems that are adequately sensitive to detect unconcentrated contaminants in wastewater treated for human consumption; however, none is available. Until such methods are developed, one must concentrate the water and use toxicological systems that are only moderately sensitive and not completely validated for this use to assess the potential health effects from exposure to complex mixtures of chemicals.

The amount of material to be used for toxicological evaluation should be adequate to conduct the Phase 1 studies described in the following text. In the event that Phase 2 and Phase 3 studies are also to be conducted, considerable quantities (i.e., gram quantities) of the concentrate are likely to be needed. If the treatment modality and the source of water have not changed, obtaining samples poses no real problem. However, the stability of the material over time must also be determined. If the composition of the concentrate deteriorates or changes with time, then the problems are further compounded (see Chapter 4). For the research (pilot-plant) phase of a study, fingerprint chemical analyses should be conducted on each batch of water concentrate to obtain a measure of the constituents. Because it will not be possible to conduct all the bioassays at the same time and because samples will be taken on a continuing basis, it is important to know how the concentrates differ in chemical composition from test to test.

Several logistical problems will arise when a toxicity testing scheme is undertaken. Studies in animals are likely to require sizable quantities (i.e., grams) of water concentrates. Either a large sample can be obtained at one time and used throughout the study, or samples can be prepared on a continuing basis. The latter sampling procedure has the advantage of being representative of chronic human water consumption patterns. For concentrates with demonstrated stability over time, concentrates taken on several occasions could be combined and the entire sample subjected to toxicological testing. This method will reduce the possibility that water samples taken at any one time are unrepresentative. The length of the interval between samplings will depend on the changing nature of the chemical composition and the degree of certainty desired. Whenever there is a major change in the supply of raw water or the treatment process, then a sample should be taken and concentrated for toxicological evaluation.

Central to the toxicity testing sequence is consideration of the extent to which the sample should be concentrated. The goal is to obtain sufficient concentration to be able to measure toxic effects while minimizing artifact formation. The potential for artifact formation from the concentration technique should be evaluated prior to dose-response testing. This may be accomplished with short-term tests (e.g., the Ames/*Salmonella* assay) at a constant dose for different concentration multiples or by measuring cytotoxicity in cell cultures. Dose-response testing would then be conducted at a concentration multiple for which no evidence of artifact formation can be shown. Because of the uncertainty in this area, the panel recommends preparation of concentrates by at least two complementary procedures and the testing of each for artifact formation. (See Chapter 4.)

Dose selection for animal testing is also of concern. It is important that the doses used be reasonable multiples (e.g., 100 to 1,000 times) of anticipated human exposure. They should be sufficient to establish an effect level, and any resultant dose-response relationship should be used to determine the comparative risks. Such studies should be performed on a comparative basis with equivalent multiples from conventional water sources. The panel considers this approach more relevant than a maximum tolerated dose (MTD) (National Cancer Institute, 1976) approach and more consistent with the reality of testing complex mixtures present at very large dilutions in the original water.

The panel recommends three phases of tiers of toxicity testing. In the following scheme for toxicological evaluation, Phase 1 includes (1) *in vitro* assessments of mutagenic and carcinogenic potential by means of microbial and mammalian cell mutation and (2) *in vivo* evaluations of acute and subchronic toxicity, teratogenicity, and clastogenicity. Phase 2 includes a 90-day subchronic study and a test for reproductive toxicity. Phase 3 is a chronic lifetime feeding study. The proposed tests are outlined in Table 5-1. Their interpretations are discussed further in Chapter 7. It should be emphasized that the proposed testing hierarchy is exploratory in nature and needs vigorous validation to ensure its overall usefulness.

TABLE 5-1 Toxicological Tests

Conventional Water	Reused Water
PHASE 1	
In *Vitro*:	
Mutagenicity	Mutagenicity
In *vitro* transformation	In *vitro* transformation
In *Vivo*:	
Acute toxicity	Acute toxicity
Teratogenicity	Teratogenicity
Short-term, repeated dose studies--14-day (includes cytogenetics assay)	Short-term, repeated dose studies--14-day (includes cytogenetics assay)
PHASE 2	
Subchronic 90-day study in at least one rodent species, preferably in two species	Subchronic 90-day study in at least one rodent species, preferably in two species
Reproductive toxicity	Reproductive toxicity
PHASE 3	
Chronic lifetime feeding study in one species of rodent	Chronic lifetime feeding study in one species of rodent

Phase I Testing

In Vitro Tests

Short-term tests, including mutagenicity and in vitro transformation models, have been used to identify mutagenic activity and to predict carcinogenic potential of chemicals from diverse environmental sources. The correlation between these tests and whole animal studies has been sufficiently high to make these valuable in screening large numbers of chemicals to indicate the need for further toxicological evaluations. The preliminary evaluation of reused water for possible health effects is expected to require testing of a relatively large number of water concentrates from pilot treatment plants because treatment processes are anticipated to be highly variable. This requirement necessitates the use of short-term tests as a part of the initial biological testing scheme.

Short-term tests can be grouped according to the specific utility of the information required. One such grouping was recommended at a workshop in San Antonio, Texas, by the Toxic Substances Control Act-Interagency Testing Committee (1979):

- point mutation in Salmonella typhimurium (Ames assay)
- gene mutation in mammalian cells such as mouse lymphoma or Chinese hamster ovary models
- in vitro transformation

These three types of assays, used as a battery, were recommended because of their correlation with rodent cancer bioassays and the high degree of confidence in a negative result.

The panel recommends that four Salmonella tester strains, (i.e, TA1535, TA1537, TA98, and TA100) be used in the Ames assay. Base-pair substitution mutations are detected with TA1535 and TA100, and frameshift mutations with TA1537 and TA98. Water concentrates should be tested directly on these strains, both with and without metabolic activation provided by rat liver microsomal enzyme preparations (e.g., S9).

To gain perspective on positive results in a prokaryotic system such as Salmonella and to detect mutagens that do not affect this system, a mammalian cell gene mutation assay is included in the battery. The mouse lymphoma model, using the thymidine kinase locus (TK+/-, TK-/-), or the Chinese hamster ovary cell model, using the hypoxanthine guanine phosphoribosyl transferase locus (HGPRT+/-, HGPRT-/-) is commonly used for this purpose.

An in vitro cell transformation assay is included as a more direct measure of carcinogenic potential and to detect compounds that do not appear to be genotoxic carcinogens. Substances that are presumed to be nongenotoxic carcinogens would not be detected in the Ames/Salmonella or mammalian cell gene mutation assays, which have a purely genetic end point. Several in vitro transformation models may be used. The BALB/3T3 and C3H/10T1/2 models are continuous cell lines derived from mice. The Syrian hamster clonal assay uses primary embryonic tissue as a cell source. Having investigators with considerable experience with such assays is more important to the successful application of these tests than the selection of one model over another.

This Phase 1 battery is proposed for the initial evaluation of reused water concentrates. However, this grouping of tests does not evaluate the potential of reused water to produce chromosome-level effects, such as breaks or translocations. For this reason, an *in vivo* cytogenetics analysis is proposed as part of the 14-day rodent study. Metaphase analysis of bone marrow cells would be performed following 7 days of dosing with administration of various levels of reused water concentrates.

The combined use of these four assays provides information on the potential of water concentrates to produce both gene or point mutations, chromosome aberrations *in vivo*, and *in vitro* transformation. Such information, evaluated in the context of other toxicological data, can provide a good initial indicator of mutagenic and carcinogenic potential of reused water. However, these tests, with the exception of the *in vivo* cytogenetic study, provide data that must be further evaluated in whole animal systems in order to extrapolate them to possible effects on human health. The whole animal testing required includes studies for oncogenicity and heritable effects on germ cells in rodent models. Information available at this time indicates that the vast majority of chemicals producing positive results in chromosome-level germ cell assays and in an *in vivo* cytogenetics assay also produces positive results using somatic bone marrow cells. Thus, results in this assay would be a good predictor of anticipated germ cell effects.

In Vivo Tests

The use of laboratory animals for toxicological testing is essential for the ultimate assessment of the possible health effects of reused water. If one is to determine the risks or the possible health hazards of repeated or continuous exposure of humans to reused water, the data obtained from similar exposures of animals should be evaluated in conjunction with the data obtained from short-term *in vitro* and *in vivo* testing for potential mutagenic and carcinogenic effects. Unlike the *in vitro* assays with bacterial or mammalian tissue or cell cultures in which effects on single systems can be tested, toxicological testing in experimental animals can identify the integrated effects of the test material on complex biological systems. Whole animal tests measure the onset and duration of action—which are dependent on such factors as absorption, distribution to critical sites, biotransformation, and excretion—during acute, subchronic, and chronic exposures (National Academy of Sciences, 1975).

The initial *in vivo* test is the determination of acute toxicity. This test is essential to provide a basis to determine acute risk and is conducted for the following reasons: to develop a profile of effects (toxicological profile), to establish a relationship between dose and observed effects, to identify target organ toxicity, to provide insight into possible mechanisms of acute toxicity, to provide adequate data to permit comparisons with other compounds evaluated acutely, and to provide data to permit determination of the LD_{50}, if needed. Although the numerical LD_{50} is often used in toxicity comparisons of compounds, this practice is not recommended unless a number of other factors are simultaneously considered. These include the potential for human exposure at high concentrations, the design and implementation of the study, method of determination of the LD_{50}, and the slope of the dose-effect relationship. Arguments offered for not conducting an LD_{50} test include (1) the large quantities of material that are necessary if the acute toxicity is low, (2) the inability of acute tests to predict cumulative effects, (3) the limited value of an LD_{50}, and (4) the uncertainty of interpreting an LD_{50} of a complex, uncharacterized mixture. Acute toxicity testing is <u>not</u> equivalent to an LD_{50} determination, and an evaluation of acute effects need not involve determination of an LD_{50}.

The acute toxicity test should be conducted in healthy young adult male and female rodents (rats and/or mice) (National Academy of Sciences, 1977b). There should be at least three doses (one dose per experimental group of five males and five females) selected for their ability to produce a range of effects and mortality, and they should be administered by gavage. If a dose of 5,000 mg/kg body weight (bw) fails to produce compound-related mortality, additional testing is not warranted. A control group, receiving only the vehicle (distilled water, if possible), should also be included. All animals should be observed frequently during the 14 days following gavage. Onset, duration, and detailed descriptions of effects and time of death must be carefully noted. All animals that succumb during the 14-day period should be necropsied. All survivors should be killed on the 14th day and necropsied. The observation period may be lengthened, depending on the effects produced.

The LD_{50} analysis, if performed, should be determined by an accepted method (e.g., Bliss, 1935; Litchfield and Wilcoxon, 1949).

Short-Term, Repeated-Dose Studies

These studies are usually conducted after acute testing. Whereas acute testing involves effects (immediate or delayed) following single-dose exposure, short-term, repeated-dose studies involve repeated administration of the test substance and observation of effects (usually delayed) due to accumulation of material in tissues or cumulative effects over a limited time. In addition, short-term, repeated-dose studies should provide evidence of target organ toxicity, the nature and development of toxicological effects, and the dose-response relationship between extent of exposure and effects produced. The data should also be used as the basis for the selection of doses for the subchronic (90-day) study in Phase 2 testing.

Short-term, repeated-dose studies should be conducted in acclimated, healthy young adult male and female (nulliparous and nonpregnant) rats or mice. There should be at least 10 males and 10 females per group, at least three dose levels per concentrate, and a control group. The control group should receive unconcentrated

conventional water, three groups should receive concentrate from
conventional water, and three should receive concentrate from reused
water. The number of doses can be increased if results are equivocal
(i.e., not dose-dependent). The limiting dose is 1,000 mg/kg bw.
The test material should be administered by gavage for 14 or 28
days. Other routes of exposure (dietary, drinking water) should also
be considered. The size of the doses used should be based on a
number of considerations, including anticipated human exposure
levels. The highest dose should produce adverse effects, but not a
significant number of fatalities; the lowest dose should not produce
evidence of toxicity. The material should be administered daily for
14 or 28 days. Observations, including changes in general appearance, behavior, and physiological functions (e.g., cardiovascular,
respiratory, and gastrointestinal functions), should be made at least
twice daily for the duration of the study. Body weight and food
consumption should be determined weekly.

Clinical observations should be made at the end of the exposure
period. These include hematological, biochemical, and cytogenetic
analyses (metaphase) of femur marrow cells and urinalyses. Hematological studies include hematocrit analysis, hemoglobin count,
erythrocyte count, total and differential leucocyte count, and
clotting potential (clotting time, prothrombin or thromboplastin
time, or platelet count). Clinical biochemical determinations
include analyses of electrolytes, proteins (total, A/G ratio),
fasting glucose, serum glutamic pyruvic transaminase (SGPT), serum
glutamic oxaloacetic transaminase (SGOT), ornithine decarboxylase,
urea nitrogen, creatinine, and bilirubin. Urinalyses should include
measures of specific gravity, reducing substances, pH, and protein
and microscopic (crystals, cells) studies.

All animals that die during the study or are killed at the end of
the observation period should be necropsied. The following organs
should be weighed and organ/body weight and organ/brain weight ratios
determined: adrenals, brain, gonads, heart, liver, kidneys, and
spleen. Histopathological examinations should be conducted on all
tissues with gross lesions or changes in size (because they are likely to be target organs) and on the adrenals, brain, gonads, heart,
liver, kidneys, and spleen of the animals receiving the highest dose
level and the control group. If differences are observed, the examinations should be extended to the appropriate tissues in the remaining
treated groups.

Cytogenetic Study

A small group of male mice or rats (5 to 10 animals) should be
included in each test and control group to evaluate the clastogenic
activity of the concentrate. At the end of a 7-day dosing period,
colcemid should be administered 2 hours before killing the animals,
femur bone marrow cells collected, and metaphase analysis performed.
Testing results to date indicate that few, if any, substances that
produce chromosome-level effects in germ cells would provide negative
results in *in vivo* cytogenetics analyses.

Teratological Testing

Teratology is the study of the causes and development of congenital malformations. Congenital malformations are generally defined as anatomical abnormalities present at birth that are lethal or that seriously interfere with normal functions. Abnormalities of this sort are the usual end points in testing for the teratogenic potential of environmental agents. At present, in vivo teratological testing in which such end points are scored is accepted as the choice for the greatest relevance. Other procedures for assessing embryologic toxicity, such as various in vitro methods or the use of functional end points or minor structural deviations, are considered less dependable than the standard in vivo methods; their general reliability rests on at least 20 years of use.

Because mice, rats, and rabbits are well characterized physiologically, economical, and convenient to use, most routine teratological testing is performed with these species. At least two species should be used to test reused water concentrates. Because of their genetic composition, random-bred (outbred) animals are usually preferred over inbred (genetically homogeneous) strains. The reasons cited for using such stocks are their greater vigor and fertility and their genetic heterogeneity--hence, their resemblance to human populations. The vigor and fertility are not compelling, and heterogeneity is probably incorrect. On the other hand, using genetically homogeneous lines of animals presents some clear advantages (Festing, 1979; Kalter, 1978).

It has usually been found that, among a small number of inbred strains randomly chosen for teratological testing, there exists a range of susceptibilities. Thus, by the concurrent use of a relatively small number of unrelated strains, two important ends can often be achieved. First, information can be obtained about the range of sensitivities to environmental agents (not only the average sensitivity) that can be present in a species; and, second, sensitive indicators will usually be observed, enabling investigators to make more realistic risk assessments (Kalter, 1981).

A recommended schedule of dosing (National Academy of Sciences, 1977b) is to administer the test material daily from the 6th through the 15th days of pregnancy in rodents and through the 18th day in rabbits, or even later (Collins 1978), so as to include the entire period of organogenesis. This regimen may be reasonable for agents that human beings are exposed to chronically, but an extended period of administration may interfere with evaluation of teratogenic potential, e.g., by causing nonspecific maternal or conceptual toxicity or by promoting or inhibiting metabolism of the test substance.

For these reasons, and also because the concentrates of substances in reused water may be in short supply, the panel recommends that the test material be administered from the 9th through the 13th days of pregnancy in rodents and from the 10th through the 14th days in rabbits because indications of potential for causing major congenital malformations may be obtained by treating the animals for relatively short intervals during the time of maximal teratological sensitivity. The test material should be administered orally, because that is the route by which humans are exposed.

More than 30 years of studies in experimental mammalian teratology have established two general facts regarding the pattern of the dose-response relationship in fetuses: (1) Almost without exception, all chemicals, if administered at sufficient levels, will have embryotoxic effects. (2) Conversely, embryotoxic substances have operational threshold dosages below which adverse effects are not observed in conventional studies (Kalter, 1968; Wilson, 1973).

The primary purpose of teratological testing, therefore, is not to discover whether or not an agent is embryotoxic, but to determine the dose-response curve that can be used to make decisions about tolerable levels of risk to human health. The usual means of achieving this goal is to determine those dosages that are significantly embryotoxic and then, by serially reducing these amounts, to arrive at doses that produce no observable effects.

In instances where quantities of the test materials are limited, as is the case with concentrates of substances in reused water, it may be necessary to proceed otherwise. An alternative procedure might be to base the dosing schedule on expected levels of human exposure and to administer multiples of these levels. This regimen was used in the only teratological study of drinking water concentrates that appears to have been conducted to date (Kavlock et al., 1979). With this procedure, realistic estimates of environmental concentrations must be made, with due recognition of variations in water composition and water consumption.

The induction of malformations in a test species is evidence of potential teratogenicity of the test material in humans. In the absence of malformations, other embryotoxic effects (increased frequency of fetal death, growth retardation, minor defects, anatomical variants) present difficulties of interpretation, because they can result from nonspecific maternal and/or fetal toxicity or perturbation of fetal homeostatic processes. Such occurrences must be evaluated individually (Kalter, 1981; Kimmel and Wilson, 1973).

Phase 2 Testing

Subchronic Toxicity Study (90 days)

This study is conducted to evaluate adverse effects of the test concentrate administered daily to experimental animals for approximately 10% of their life span. Such studies are meant to identify target organs, to determine the lowest observed adverse effect level and the highest no observed adverse effect level, and to provide information on cumulation (of effects and of levels in tissues); to evaluate effects on reproductive performance including gonadal function, estrous cycles, mating behavior, conception, parturition, lactation, weaning, and development; and to provide information on neonatal and adult morbidity and mortality.

For this study, the first generation (F_{1a}) of parents (F_0) exposed to the test materials is used. The parents are exposed from weaning until 80-100 days of age, when they are mated, and the exposure continues throughout gestation and lactation. Rats or mice are generally used. The duration of exposure for the F_{1a} generation is continuous from conception until 90 days past weaning. Nonrodents should be exposed for 1 year. There is some experimental evidence that toxicity observed in a 90-day toxicity test may correlate well with the results (with the exception of cancer) of a lifetime (i.e., chronic) study in rats and dogs (Weil and McCollister, 1963). At least 20 males and 20 female rodents per group should be used. A control group should receive unconcentrated, conventional water; at least three dose groups should receive concentrates of conventional water; and at least three dose groups should receive concentrates of reused water. The group receiving the highest dose should exhibit adverse effects, but a low incidence of mortality. Ideally, the group receiving the lowest dose should show no evidence of an adverse effect. The test materials should be administered daily by gavage for 90 consecutive days (or 1 year for nonrodents). Alternatives to gavage (e.g., dietary or drinking water exposures) should be considered.

Animals should be observed at least twice daily for general appearance, behavior, and physiological functions (e.g., cardiovascular, respiratory, and gastrointestinal functions). The time of onset, the duration, and the extent of effects should be carefully noted. Food consumption (and water consumption when indicated), body weights, and detailed physical findings should be determined weekly. Clinical observations should include: ophthalmological (at onset and termination of the study), hematological, biochemical, and urinary examinations. Hematological studies include hematocrit analysis, hemoglobin and erythrocyte counts, total and differential leucocyte counts, and measures of clotting potential (clotting time, prothrombin or thromboplastin time, platelet count). Clinical biochemical determinations should include assessment of electrolyte balance, carbohydrate metabolism, liver and kidney function, and other tests as determined by the observations made. These should include analyses of electrolytes, fasting glucose, SGPT, SGOT, serum alkaline phosphatase (SAP), bilirubin, proteins (total and A/G ratio), blood urea nitrogen (BUN), creatinine, and ornithine decarboxylase. Urinalyses should include measures of specific gravity, reducing substances, pH, and protein and microscopic studies (crystals, cells). Hematological, biochemical, and urinary evaluations should be conducted at 28 and 90 days. All animals that die during the treatment period or are sacrificed at the end should be necropsied. The following organs should be weighed and organ/body weight and organ/brain weight ratios determined: adrenals, brain, gonads, heart, liver, kidneys, and spleen. The following should be removed and preserved for histopathological examination: all gross lesions; brain (three major areas), spinal cord (three levels), and peripheral nerves; pituitary, thyroid, parathyroid, and adrenal glands; thymus and pancreas; gonads, or uterus; trachea and lungs; salivary glands; stomach, duodenum, jejunum, ileum, cecum, colon, rectum, liver, bladder, and spleen; mammary glands; skeletal muscles; sternum, femur, and marrow; and eye and lachrymal glands.

Histopathological examinations should be conducted on all tissues from control animals on all gross lesions and on target organs (if identified from animals) receiving the highest dose. The histopathological examination should be extended to include appropriate tissues in other treated groups if significant findings are observed in the group receiving the highest dose.

Phase 3 Testing

Phase 3 consists of a combination chronic toxicity and carcinogenicity study, which may be needed after the 90-day study if it becomes necessary to evaluate potential human health hazards, particularly the carcinogenic potential and other effects associated with lifetime exposure. Long-term chronic tests should be conducted, primarily on the premise that effects of lifetime ingestion by humans cannot be predicted from tests conducted in a short-lived animal for periods considerably less than a lifetime. In conventional toxicological testing, 24 to 30 months has been considered the average lifespan for the rat and 18 to 24 months for the mouse.

For testing the health hazards of reused water, the primary concerns are the effects of exposure to small amounts over a long period. Even if the results are similar to those of the subchronic tests, it is at times necessary to determine if these tests would show increased morbidity or mortality by the end of a lifetime. Ultimately, the lifetime study could be used to estimate the risk to human health presented by chronic ingestion of reused water.

For this study, the F_{1a} generation of parents exposed to the test materials should be used. The duration of exposure of the F_{1a} generation should be continuous from conception until 24 to 30 months of age. The protocol described for the 90-day study should also be used for this study. The ophthalmological examinations should be conducted at the beginning of the study and at 6, 12, and 24 months. The hematological, biochemical, and urinary evaluations should be conducted at 3, 6, 12, 18, and 24 months. The test material can be administered by gavage, in the diet, or in drinking water (preferred). At least 50 males and 50 females per dose level should be used. There should be a control group and at least three test groups. The doses to be used should be based on data obtained in the 90-day study on an appropriate multiple of the maximum anticipated human consumption (e.g., 100 or 1,000 times greater).

Practical Considerations for Chronic Testing

The lifetime feeding study should provide the definitive test of oncogenicity and other chronic effects of renovated water. A number of practical requirements limit the scope of the test with regard to dose and concentration multiples. To illustrate these limitations, some hypothetical assumptions concerning the concentrates are outlined in Table 5-2. The amounts of solutes in the unconcentrated water must be considered prior to concentration so approximately equimolar concentrates may be prepared. A discussion of these assumptions and limitations follows.

TABLE 5-2 Assumed Solute Concentrations of Conventional and Renovated Water Concentrate

Sample	Concentration (mg/liter)	
	Total Organics	Total Solids
Conventional Water		
Unconcentrated	3	300
Concentrated (100 times)	300	30,000
Renovated Water		
Unconcentrated	6	600
Concentrated (50 times)	300	30,000

Choice of Maximum Dose to Test Animals

If given as drinking water, the maximum average volume is 25 ml/rat/day. Table 5-3 lists the daily doses for rats. Thus, the relative daily dose ratio is:

$$\frac{\text{Human equivalent to maximum dose to rat}}{\text{Current human intake}} = \frac{2{,}625 \text{ mg/kg}}{6 \text{ mg/kg}} = 437$$

or approximately 450 times average human daily intake.

TABLE 5-3 Suggested Daily Dose of Drinking Water to Rats

Substance	Dose (mg/rat[a])	(mg/kg[b])	Equivalent Human Dose[c] (mg/70 kg)	Current Human Intake[d] (mg/70 kg)
Total organics	7.5	37.5	2,625	6
Total solids	750	3,750	262,500	600

[a] 25 ml x 0.3 mg/ml.
[b] $7.5 \times \frac{1{,}000}{200} = 37.5$; assumes 200 g body weight.
[c] 37.5 x 70 = 2,625.
[d] Assumes 2 liter/day of conventional water per 70-kg adult.

TABLE 5-4 Minimum Water Requirements for Concentrates

Test Solution	Dose per Rat (ml)	Total Volume per 100 Rats (ml)	Minimum Volume of Original Water Needed (liters)
Conventional water concentrate	25	2,500	250
Reused water concentrate	25	2,500	125

The minimum volume of water required for tests in animals using three groups of 50 males and 50 females plus a control group gavaged with conventional water concentrate and reused water concentrate at a volume of 25 ml/rat is summarized in Table 5-4.

Removal of Electrolytes Before Testing

Inorganic elements normally found in drinking water are listed in Table 5-5, along with the average daily intake for the adult rat and the minimal or optimal daily intake.

The actual dose of total solids after concentration is expected to be 750 mg/rat, as compared to 466 mg/rat for the sum of the ions listed. A factor of less than twice the daily average intake of total inorganic compounds in water should not place an excessive physiological stress on the animal. Potassium is the most toxic element. However, it appears unlikely that the additional intake of this element would constitute a large fraction of total inorganic solids in water.

There may be reasons to remove the inorganic solids; for example, to avoid precipitates that might remove some of the organic constituents by adsorption.

Sensitivity of the Chronic Toxicity Test--Statistical Considerations

Because of costs involved, relatively small numbers of rodents are used for each dose level in 2-year chronic studies. Typically, only 50 animals per dose are used. This limits the size of differences in tumor rates that can be detected. Differences in tumor rates that can be detected with high probability (approximately 90%) are listed in the Table 5-6 for animals fed either reused or conventional water concentrates. Differences that can be detected with 100 animals per dose group are included to show the gain from increasing the number of animals. For example, with 50 animals per group, if the tumor incidence of animals administered concentrate from conventional water

TABLE 5-5 Average Daily Intake by Rats of Inorganic Elements in Water

Inorganic Element	Average Daily Intake per Adult Rat[a] (mg)	Estimated Optimal[b] or Minimal Requirement (mg)
Calcium	121	45
Chloride	63	5
Iron	2.4	0.25
Magnesium	28	0.8
Manganese	0.85	0.8
Phosphorus	94	40
Potassium	116	15 male
		8 female
Sodium	40	40
Zinc	0.7	0.04
TOTAL	466	143

[a]Ralston Purina Co., 1980. The average intake of solid food is assumed to be 13.5 g/adult rat.
[b]Farris and Griffith, 1967.

is 5%, then the true tumor rate resulting from renovated water would have to be about 25% or more in order to have a high probability (approximately 90%) of detecting a statistically significant increase in tumor incidence for animals exposed to renovated water. If the difference in tumor rates between the two types of water is less, then there will be a lower probability of detecting this difference.

TABLE 5-6 Differences in Tumor Rates Detectable with Approximately a 90% Probability Using a One-Sided Test with Type I Error = 0.05

Differences in Tumor Rates (%)			
50 Animals/Dose		100 Animals/Dose	
Conventional Water	Reused Water	Conventional Water	Reused Water
5	25	5	20
10	35	10	25
20	50	20	40

TABLE 5-7 Approximate 95% Confidence Limits on Relative Risk[a]

Animals with Tumors in Both Water Types (%)	50 Animals/Group		100 Animals/Group	
	Lower Limit	Upper Limit	Lower Limit	Upper Limit
10	0.2	4.0	0.4	2.6
30	0.5	2.0	0.6	1.6
50	0.6	1.6	0.7	1.3

[a]From Thomas and Gart, 1977.

For the given dose levels, the proportions of animals with tumors will be described for exposures to concentrate from both renovated and conventional water. Table 5-7 gives approximate 95% confidence limits on the relative risk (ratio of tumor incidence with renovated water to the tumor incidence with conventional water) when the observed tumor incidence is the same in both experimental groups. For example, if 10% (5 out of 50 animals) of the animals develop tumors with both the renovated and conventional water, the observed risk is the same. The estimated relative risk is 1; the true relative risk may vary from 0.2 to 4.0. That is, with 95% confidence, the true tumor rate from renovated water may vary from 20% of the tumor rate with conventional water up to 4 times greater when 5 of 50 animals in each group have observable tumors.

If there are no differences in responses between male and female animals, it may be possible to pool the data from both sexes to increase the statistical sensitivity. However, it is doubtful that even a pooled analysis could reliably distinguish between, for example, a 5% response frequency from concentrates of conventionally treated water and a 10% response frequency from concentrates of reused water. The above limitations, as well as others, need to be recognized in the design of chronic tests. Appendix B provides further statistical details on sampling.

REFERENCES

Army Corps of Engineers. 1981. Operation, Maintenance and Performance Evaluation of the Experimental Estuary Water Treatment Plant--First Progress Report, May 1980-March 1981. James M. Montgomery Consulting Engineers, Inc., Pasadena, Calif. Various paging.

Bliss, C.L. 1935. The calculation of the dose-mortality curve. Ann. Appl. Biol. 22:134-167.

Cheh, A.M., J. Skochdopole, C. Heilig, P.M. Koski, and L. Cole. 1980. Destruction of direct acting mutagens in drinking water by nucleophiles: Implications regarding mutagen identification and mutagen elimination from drinking water. Pp. 803-816 in R.L. Jolley, W.A. Brungs, R.B. Cumming, and V.A. Jacobs, eds. Water Chlorination: Environmental Impact and Health Effects, Vol. 3. Ann Arbor Science, Ann Arbor, Mich.

Chernoff, N., E. Rogers, B. Carber, R. Kavlock, and E. Gray. 1979. Fetotoxic potential of municipal drinking water in the mouse. Teratology 19:165-169.

Collins, T.F.X. 1978. Reproduction and teratology guidelines: Review of deliberations by the National Toxicology Advisory Committee's Reproduction Panel. J. Environ. Pathol. Toxicol. 2:141-147.

Crump, K.S., and H.A. Guess. 1980. Drinking water and cancer: Review of recent findings and assessment of risks. Council on Environmental Quality, Washington, D.C. 107 pp.

de Greef, E., J.C. Morris, C.F. van Kreijl, and C.F.H. Morra. 1980. Health effects in the chemical oxidation of polluted waters. Pp. 913-924 in R.L. Jolley, W.A. Brungs, and R.B. Cummings, eds. Water Chlorination: Environmental Impact and Health Effects, Vol. 3. Ann Arbor Science, Ann Arbor, Mich.

Doll, R., and R. Peto. 1981. The causes of cancer: Quantitative estimates of avoidable risk of cancer in the United States today. J. Natl. Cancer Inst. 66:1191-1308.

Environmental Protection Agency. 1979a. Interim Primary Drinking Water Regulations; Amendments. July 19, 1979. Fed. Reg. 44:42246-42259.

Environmental Protection Agency. 1979b. National Secondary Drinking Water Regulations. July 19, 1979. Fed. Reg. 44:42195-42202.

Farris, F.J., and J. O. Griffith. 1967. The Rat in Laboratory Investigation. Hafner Pub. Co., New York, pp. 68-82.

Festing, M.F.W. 1979. Properties of inbred strains and outbred stocks, with special reference to toxicity testing. J. Toxicol. Environ. Health 5:53-68.

Ford, J.O., and M.A. Pereira. 1980. Short-term _in vivo_ initiation/promotion bioassay for hepatocarcinogens. J. Environ. Pathol. Toxicol. 4:39-46.

Frerichs, R.R., K.P. Satin, and E.M. Sloss. 1981. Water re-use--It's epidemiologic impact, Los Angeles County, 1969-1971. Regents of the University of California, Los Angeles, Calif. 128 pp.

Grabow, W.O.K., and M. Isaacson. 1978. Microbiological quality and epidemiological aspects of reclaimed water. Prog. Water Technol. 10:329-335.

Grabow, W.O.K., P.G. Van Rossum, N.A. Grabow, and R. Denkhaus. 1981. Relationship of the raw water quality to mutagens detectable by the Ames Salmonella/microsome assay in a drinking-water supply. Water Res. 15:1037-1043.

Gruener, N. 1978. Evaluation of Toxic Effects of Organic Contaminants in Recycled Water. EPA 600/1-78-068. Health Effects Research Laboratory, Office of Research and Development, U.S. Environmental Protection Agency, Cincinnati, Ohio. 99 pp.

Gruener, N. 1979. Biological evaluation of toxic effects of organic contaminants in concentrated recycled water. Pp. 2187-2195 in Proc. Water Reuse Symp., Vol. 3. American Water Works Research Foundation, Denver, Colo.

Gruener, N., and M.P. Lockwood. 1980. Mutagenic activity in drinking water. Am. J. Publ. Health 70(3):276-278.

Hattingh, W.H.J. 1978. Health aspects. Pp. 259-283 in Manual for Water Renovation and Reclamation. Technical Guide K42. National Institute for Water Research, Council for Scientific and Industrial Research, Pretoria, Republic of South Africa.

Helms, C.T., C.C. Sigman, S. Malko, D.L. Atkinson, J. Jaffer, P.A. Sullivan, K.L. Thompson, E. M. Knowlton, H.F. Kraybill, J. Hushon, D.P. Thoman, R.S. Clerman, and N. Barr. 1981. Biorefractories in Water. Evaluation of Classification of the Potential Carcinogenicity and Mutagenicity of Chemical Biorefractories Identified in Drinking Water. U.S. Department of Health and Human Services, National Cancer Institute, Washington, D.C. 39 pp.

Kalter, H. 1968. Teratology of the Central Nervous System: Induced and Spontaneous Malformations of Laboratory, Agricultural and Domestic Mammals. University of Chicago Press, Chicago. 483 pp.

Kalter, H. 1978. Structure and uses of genetically homogeneous lines of animals. Pp. 155-190 in J.C. Wilson and F.C. Fraser, eds. Handbook of Teratology, Vol. 4. Research Procedures and Data Analysis. Plenum Press, New York.

Kalter, H. 1981. Dose-response studies with genetically homogeneous lines of mice as a teratology testing and risk-assessment procedure. Teratology 24:79-86.

Kavlock, R., N. Chernoff, B. Carver, and F. Kopfler. 1979. Teratology studies in mice exposed to municipal drinking-water concentrates during organogenesis. Fd. Cosmet. Toxicol. 17:343-347.

Kimmel, C.A., and J. G. Wilson. 1973. Skeletal deviations in rats: Malformations or variations? Teratology 8:309-315.

Lang, D.R., H. Kurzepa, M.S. Cole, and J.C. Loper. 1980. Malignant transformation of BALB/3T3 cells by residue organic mixtures from drinking water. J. Environ. Pathol. Toxicol. 4:41-54.

Litchfield, J.T., Jr., and F. Wilcoxon. 1949. Simplified method of evaluating dose-effect experiments. J. Pharmacol. 96:99-113.

Loper, J.C. 1980. Mutagenic effects of organic compounds in drinking water. Mut. Res. 76:241-268

McCann, J., E. Choi, E. Yamasasaki, and B.N. Ames. 1975. Detection of carcinogens as mutagens in the Salmonella/microsome test: Assay of 300 chemicals. Proc. Natl. Acad. Sci. USA 72:5135-5139.

McCarty, P.L., J. Kissel, T. Everhart, R.C. Cooper, and C. Leong. 1980a. Mutagenic Activity and Chemical Characterization for the Palo Alto Wastewater Reclamation and Groundwater Injection

Facility. Tech. Rep. No. 254. Department of Civil Engineering, Stanford University, Stanford, Calif. 65 pp.

McCarty, P.L., M. Reinhard, J. Graydon, J. Schreiner, K. Sutherland, T. Everhart, and D.G. Argo. 1980b. Advanced Treatment for Wastewater Reclamation at Water Factory 21. Tech. Rep. No. 236, Grant No. EPA-S-803873. Department of Civil Engineering, Stanford University, Stanford, Calif. 149 pp.

McKinney, J.D., R.R. Maurer, J.R. Hass, and R.O. Thomas. 1976. Possible factors in the drinking water of lab animals causing reproductive failure. Pp. 417-432 in L.H. Keith, ed. Identification and Analysis of Organic Pollutants in Water. Ann Arbor Science, Ann Arbor, Mich.

National Academy of Sciences. 1975. Principles for Evaluating the Environment. Committee for the Working Conference on Principles of Protocols for Evaluationg Chemicals in the Environment, National Research Council. National Academy of Sciences, Washington, D.C. 454 pp.

National Academy of Sciences. 1977a. Drinking Water and Health. Safe Drinking Water Committee, Advisory Center on Toxicology, Assembly of Life Sciences, National Research Council. National Academy of Sciences, Washington, D.C. 939 pp.

National Academy of Sciences. 1977b. Principles and Procedures for Evaluating the Toxicity of Household Products. Committee for the Revision of NAS Publication 1138. Committee on Toxicology, Assembly of Life Sciences, National Research Council. National Academy of Sciences, Washington, D.C. 199 pp.

National Academy of Sciences. 1980a. Drinking Water and Health, Vol. 3. Safe Drinking Water Committee, Board on Toxicology and Environmenal Health Hazards, Assembly of Life Sciences, National Research Council. National Academy Press, Washington, D.C. 415 pp.

National Academy of Sciences. 1980b. Principles of Toxicological Interactions Associated with Multiple Chemical Exposures. Panel on Evaluation of Hazards Associated with Maritime Personnel Exposed to Multiple Cargo Vapors, Board on Toxicology and Environmental Health Hazards, Assembly of Life Sciences, and Committee on Maritime Hazardous Materials, Commission on Sociotechnical Systems, National Research Council. National Academy Press, Washington, D.C. Various paging.

National Academy of Sciences. 1982. Drinking Water and Health, Vol. 4. Safe Drinking Water Committee, Board on Toxicology and Environmental Health Hazards, Assembly of Life Sciences, National Research Council, National Academy Press, Washington, D.C. 299 pp.

National Cancer Institute. 1976. Guidelines for Carcinogenic Bioassay in Small Rodents. DHEW Pub. No. (NIH) 76-801. Carcinogenic Testing Program, National Cancer Institute, National Institutes of Health, Bethesda, Md.

Nellor, M.H., R.B. Baird, and W.E. Garrison. 1981. Health Effects of Water Reuse by Groundwater Recharge. County Sanitation Districts of Los Angeles County, Whittier, Calif. 31 pp.

Pelon, W, T.W. Beasley, and D.E. Lesley. 1980. Transformation of the mouse clonal cell line R846-DP8 by Mississippi River, raw, and finished water samples from Southeastern Louisiana. Environ. Sci. Technol. 14:723-726.

Ralston Purina Co. 1980. Specifications for Certified Rodent Chow No. 5002. Ralston Purina Co., St. Louis, Mo.

Robinson, M., J.W. Glass, D. Cmehil, R.J. Bull, and J.G. Orthoefer. In press. Initiating and promoting activity of chemicals isolated from drinking waters in the SENCAR mouse--A five city survey. Second Symposium on Application of Short-Term Bioassays in the Fractionation and Analysis of Complex Environmental Mixtures. Health Effects Research Laboratory, U.S. Environmental Protection Agency, Cincinnati, Ohio.

Staples, R.E., W.C. Worthy, and T.A. Marks. 1979. Influence of drinking water--tap versus purified--on embryo and fetal development in mice. Teratology 19:237-243.

Tabor, M.W., J.C. Loper, and K. Barone. 1980. Analytical procedures for fractionating non-volatile mutagenic components from drinking water concentrates. Pp. 899-912 in R.L. Jolley, W.A. Brungs, and R.B. Cumming, eds. Water Chlorination: Environmental Impact and Health Effects, Vol. 3. Ann Arbor Science, Ann Arbor, Mich.

Tardiff, R.G., and M. Deinzer. 1973. Toxicity of organic compounds in drinking water. Pp. 23-37 in V. L. Snoeyink and M.F. Whelan, eds. Proceedings, Fifteenth Water Quality Conference--Organic Matter in Water Supplies: Occurrence, Significance, and Control. University of Illinois, Champaign-Urbana, Ill.

Thomas, D.G., and J.J. Gart. 1977. A table of exact confidence limits for differences and ratios of two proportions and their odds ratios. J. Am. Stat. Assoc. 72:73-76.

Toxic Substances Control Act--Interagency Testing Committee. 1979. Scoring Chemicals for Health and Ecological Effects Testing. Proceedings of a Workshop in San Antonio, Texas, February 25-28. Enviro Control, Inc., Rockville, Md.

Van Rensburg, S.J., W.H.J. Hattingh, M.L. Siebert, and N.P.J. Kriek. 1977. Biological testing of water reclaimed from purified sewage effluents. Presented at an International Conference on Advanced Treatment and Reclamation of Wastewater, 13-17 June.

Weil, C.S., and D.D. McCollister. 1963. Relationship between short- and long-term feeding studies in designing an effective toxicity test. J. Agric. Food Chem. 11:486-491.

Wilson, J.G. 1973. Environment and Birth Defects. Academic Press, New York. 305 pp.

6
Strategies for Assessing and Monitoring Water Quality for Human Exposure

Low-risk drinking water from high-quality water sources is not guaranteed by (current) drinking water standards. As scientific knowledge increases, these standards are subject to further review; they should never be interpreted as inclusive of all risk from all compounds present in tap water. For this reason, existing standards assume that the highest quality source will be used for water for human consumption.

The panel recommends demands that even treated wastewater be initially regarded as contaminated and that prior to human consumption every effort be expended to establish its comparative safety. Greatly increased importance is, therefore, attached to the recognition of all chemical and biological agents present and to the identification of those that may be particularly hazardous to public health. This task presents a considerable challenge for the following reasons:

- Our present measurement techniques are not capable of complete chemical identification.
- Health effects testing is based on responses in nonhuman systems; in wastewater treatment systems this type of testing is incomplete and largely untried.

In this context, the panel has directed its efforts toward analyzing those assessment and monitoring strategies that experience suggests have the greatest predictive value for determining potability. The issues of reliability of wastewater treatment systems, of chemical and biological monitoring (including the use of surrogate tests), and of toxicity will be addressed separately.

RELIABILITY OF WASTEWATER TREATMENT SYSTEMS

Much of the published data regarding wastewater treatment comes from a relatively small number of advanced wastewater treatment plants. Studies of these systems suggest that metals can be removed with varying efficiencies, although considerable variability exists in the quality of effluent from plant to plant.

With regard to inorganic compounds, especially metals, several issues are of particular concern:

- Some potentially toxic metals are known to be present, including arsenic, cadmium, chromium, copper, lead, mercury, selenium, and silver.
- The oxidation state, the nature of organometallic interactions, and the degree to which these are affected by various advanced waste treatment processes are not completely understood.
- Results of studies have suggested that it is not possible to predict accurately the inorganic composition of influents and effluents. Variations at specific sites over time must be evaluated on a case-by-case basis.

Therefore, great emphasis is placed on monitoring frequency and in-plant quality control. Hart (1978) noted that treatment reliability can be regarded as the capacity of a plant to remove contaminants continuously and to meet standards set for the final water. He suggests the following to achieve reliability:

- determination of the occurrence, concentration, and types of toxic inorganic chemical compounds, pathogenic microorganisms, and hazardous organic chemical compounds to be expected in the raw water supply;
- identification of critical parameters associated with each of these compounds and determination of the correlation between each parameter and the substance or organism;
- development of a control strategy to maintain plant performance;
- establishment of a minimum of two contaminant barriers for each of the substances or organisms;
- establishment of a monitoring system whereby all parameters are monitored with a frequency that is truly representative of the product quality; and
- maintenance of permanent records to evaluate plant reliability and product quality.

Contaminant barriers must be established in the context of the specific substances and the individual processes that will remove them effectively. The barriers for the Stander Plant at Windhoek are listed in Table 6-1. Hart noted that in this system various treatment processes, including lime treatment, chlorination, and activated carbon adsorption, serve as barriers against different concentrations of organic contaminants. The overall removal efficiency is 99%, except for volatile organohalogen compounds.

At Water Factory 21, treatment plant reliability is enhanced by the ability to treat water at constant flow rates and to shut down operation whenever influent wastewater characteristics are poor or an equipment failure occurs. This plant is also shut down at various periods for routine maintenance. Even so, the wastewater characteristics vary considerably from day to day.

TABLE 6-1 Contaminant Barriers in the Stander Plant[a]

Inorganic Compounds ($\mu g/dm^3$)

Sampling Point	Ba	NO_3 (as N)	CN	Zn	As	Cd
Biofilter humus tank effluent	<250			62	<10	<10
Lime treatment						
Quality equalization						
Ammonia stripping						
Bicarbonation						
Sand filtration						
Chlorine disinfection						
Active carbon adsorption						
Final treatment	<250			<25	<10	<10
Standards for drinking water						
South African Bureau of Standards	NS[b]	10,000	10	5,000	50	50
World Health Organization	NS	10,000	50	5,000	50	10
U.S. Environmental Protection Agency	1,000	10,000	NS	NS	50	10
Federal Republic of Germany (1975)	NS	20,000	50	2,000	40	6

[a]From Hart, 1978.
[b]NS = not specified.

TABLE 6-1 Continued

Sampling Point	Inorganic Compounds ($\mu g/dm^3$)				
	Cr	Pb	Hg	Se	Ag
Biofilter humus tank effluent	<50	<100	1.5		<25
Lime treatment					
Quality equalization					
Ammonia stripping					
Bicarbonation					
Sand filtration					
Chlorine disinfection					
Active carbon adsorption					
Final treatment	<50	<100	<1		<25
Standards for drinking water					
South African Bureau of Standards	50	150	NS	NS	NS
World Health Organization	NS	100	1	10	NS
U.S. Environmental Protection Agency	50	50	2	10	50
Federal Republic of Germany (1975)	50	40	4	8	NS

TABLE 6-1 Continued

Organic Compounds

Sampling Point	PAH[c] (Avg. No. of Peaks)		CH[c]	VOH[c]	TOH[c]
	>20 ng/dm^3	2 to 20 ng/dm^3	(mg/dm^3)		
Biofilter humus tank effluent					
Lime treatment					
Quality equalization					
Ammonia stripping					
Bicarbonation					
Sand filtration	4.2	2.6	1.07	43.2	23.9
Chlorine disinfection	0.6	0.8	1.18	90.0	23.2
Active carbon adsorption	1.0	1.4	0.45	1.8	5.6
Final treatment	0.4	0.8	0.44	10.7	7.1
Standards for drinking water					
South African Bureau of Standards	NS				
World Health Organization	NS				
U.S. Environmental Protection Agency	NS		1,092		
Federal Republic of Germany (1975)	250 ng/dm^3 as C				

[c] PAH--Polynuclear aromatic hydrocarbons; CH--Chlorinated hydrocarbons; VOH--Volatile organic hydrocarbons; TOH--Total organic hydrocarbons.

TABLE 6-1 Continued

Sampling Point	Geometric Mean of Microorganisms			
	Total Count Plate per cm^3	Total Coliforms per 100 cm^3	Clostridium perfringens per 100 cm^3	Enteric Viruses per 100 dm^3
Biofilter humus tank effluent	65,000	150,000	2,000	600
Lime treatment	940	200	--	--
Quality equalization	940	200	--	--
Ammonia stripping	--			
Bicarbonation	--			
Sand filtration	2,500	250		
Chlorine disinfection	6	0.11[d]	0.44[d]	0
Active carbon adsorption	--	--		
Final treatment	4	0.09[d]	0.05[d]	0
Standards for drinking water				
South African Bureau of Standards				
World Health Organization				
U.S. Environmental Protection Agency	NS	1	NS	0
Federal Republic of Germany (1975)				

[d]Arithmetic mean (100/cm^3).

MONITORING AND ASSESSING EXPOSURES

In evaluating the possible exposure of humans to chemicals in treated wastewater to be used for potable purposes, several questions must be considered. First, what constituents can be expected to be present, and at approximately what concentrations? Second, are any of these likely to be present at concentrations that could be harmful? And, finally, what analytical measurements should be undertaken, both before instituting reuse and during actual reuse? In the latter case, this procedure would, in effect, constitute a monitoring strategy.

One framework for deciding on those constituents that are of public health concern was developed by Englande and Reimers (1979). They defined an "exceedance" ratio (ER) as the number of samples surpassing potable water quality, divided by the total number of samples evaluated. They then applied this formula to several constituents analyzed in effluents from five advanced wastewater treatment systems. Table 6-2 presents the data for those constituents with positive ER values. As shown there, ammonia (not covered by U.S. drinking water standards) generally had a high value; chromium, arsenic, and iron also appear frequently in the table. The facilities that employed lime treatment reported the highest and most consistent reductions in heavy metal concentrations. Although these ER's do not in themselves constitute a decisionmaking tool for monitoring, they do point out some of the consistent problems and may help focus on those that should receive high priority in developing such schemes. That is, experience from the various studies of advanced wastewater treatment systems and their collective results should be examined when similar systems are being planned.

Many constituents will follow a log-normal time distribution, as discussed in Chapter 3 regarding Water Factory 21. Thus, based on actual experience, some constituents will show wide distributions of concentrations; others, more narrow ones. If the distribution of a given contaminant is likely to be wide or if the concentrations on occasion approach levels of concern to health, then the constituent becomes a logical candidate for monitoring.

The next question that arises is, what action should be taken when specific, high concentrations are reached? Action may be required when a level related to a potentially adverse health effect is attained. A constituent of concern--primarily from the effect of chronic (e.g., lifetime) exposure--might be permitted by the responsible agency occasionally to exceed the existing standard, if any. In such a case, the actual time distribution for the constituent should be used to predict the fraction of time (and its uncertainty range) that the concentration is likely to exceed some predetermined value. Reaching that time limit should signal the need for additional sampling, and judgments must be made as to whether concentrations have increased to the point that human exposure has become unacceptably high.

The final question is, which are the most useful and cost-effective set of chemical constituents that should be monitored regularly, and how frequently should they be monitored? For example, should one monitor simply those chemicals for which maximum contaminant levels (MCL's) are specified in the Environmental Protection

Agency's (EPA's) National Interim Primary Drinking Water Regulations? If a pilot plant shows that the renovated product is highly unlikely to have any such constituents at concentrations that are unusual in relation to the experience of currently acceptable public water supplies using "natural" sources, then there may be no reason to monitor these constituents more stringently in the renovation system. This conclusion may not be acceptable, however, if there is a judgment that there is a greater likelihood of failure of the treatment chain.

If, on the contrary, preliminary studies show the presence of unusually higher concentrations of chemicals of health significance (in comparison with those normally encountered in public water supplies), then proper monitoring strategies must be developed and implemented. Such an approach is predicated on the assumption that a renovated wastewater supply for potable use need not intrinsically constitute an unusual hazard. Therefore, in preliminary studies, a preliminary survey of all urban and industrial inputs to the wastewater collection system is necessary.

INDICATORS AND CHEMICAL CONTAMINATION

The complexity of the chemical and biological composition of wastewater presents two distinctly different analytical problems that must be considered in the design of monitoring programs. First, there is a need to measure routinely those collective parameters, e.g., total organic carbon (TOC) and total organic halogen (TOX), that are indicative of concentrations of groups of substances. However, total concentration values, such as TOC, may not indicate important concentration changes, particularly those of hazardous organic constituents. Indeed, only a small fraction of TOC has ever been associated with specific chemical structures, even though identified compounds number in the thousands. Another problem is related to the relative sophistication of modern gas chromatography and mass spectrometry (GC/MS) techniques, the data from which are subject to misinterpretation. Each of these problems is discussed below.

Nonspecific Analysis

Nonspecific analyses comprise a class of measurements, indicative usually of more than one and often many constituents being measured in the system. Examples of such measurements used in water or wastewater analysis are shown in Table 6-3. These measurements can be useful in monitoring wastewater renovation systems for the same reasons that they are currently used in water and wastewater analysis:

- Many of them are readily automated, rapid, or capable of being used in on-line monitoring.
- They may be considered as contamination indices.
   Technicians may require less training and expertise than do those performing analyses for many specific chemicals, especially organic compounds.

TABLE 6-2 Potable Water Quality Ratios (ER) for Advanced Wastewater Treatment Plant Spot Survey[a]

LAKE TAHOE[b]

Substance	Median (per liter)		ER
NH_3-N	15	mg	13/13
Se	16	µg	13/13
Phenol	5	µg	7/12
CCE[e]	650	µg	3/8
Fe	140	µg	3/13
As	<10	µg	2/13
Cd	<1	µg	1/13
TDS[f]	414	mg	1/13

BLUE PLAINS[c]

System 1

Substance	Median (per liter)		ER
TDS	524	mg	3/3
CCE	1,300	µg	2/2
NH_3-N	0.65	mg	2/3
Phenol	<5	µg	1/2
Cr	19	µg	1/3
Pb	19	µg	1/3

DALLAS[g]

Substance	Median (per liter)		ER
Phenol	7.2	µg	6/8
Se	35	µg	5/8
TDS	478	mg	3/8
As	21	µg	2/7
NH_3-N	0.08	mg	2/8
CCE	200	µg	1/7
Cr	4	µg	1/8
Fe	93	µg	1/8

POMONA[h]

System 1

Substance	Median (per liter)		ER
TDS	513	mg	7/9
NH_3-N	2.0	mg	5/9
Se	<10	µg	5/9
CCE	650	µg	4/8
Phenol	<5	µg	3/9
Cr	30	µg	2/9
NO_3-N	4.5	mg	2/9
Fe	7.9	µg	2/9
Hg	0.31	µg	2/9
Mn	0.0	µg	1/9

[a]From Englande and Reimers, 1979.
[b]Lake Tahoe--trickling filter (TF), nitrifying activated sludge (NAS), high lime coagulation (HLC) and clarification (C), filtration (F), carbon adsorption (CA).
[c]Blue Plains: System 1--Low lime clarification (LLC), breakpoint chlorination (BC), CA, dual media filtration. System 2--LLC, dispersed growth nitrification, denitrification, CA, mixed media filtration.
[d]Orange County--TF, HLC, and C. Ammonia stripping, F, activated carbon (AC).
[e]CCE = carbon chloroform extract.
[f]TDS = total dissolved solids.
[g]Dallas--Activated sludge (AS), HLC and C, F, CA.
[h]Pomona: System 1--AS, AC. System 2--AS, AC, chlorination, AC. System 3--AS, AC, ozonation, AC.

ORANGE COUNTY[d]

System 2 Substance	Median (per liter)		ER
Se	10	µg	1/6
NO_3-N	0.5	mg	1/6
Phenol	<5	µg	1/6

Substance	Median (per liter)		ER
NH_3-N	14.2	mg	6/6
TDS	937	mg	6/6
Cr	57	µg	5/6
NO_3-N	11.3	mg	5/6
Phenol	<5	µg	1/6

System 2 Substance	Median (per liter)		ER
NH_3-N	7.0	mg	8/9
TDS	538	mg	6/9
Phenol	<5	µg	3/9
NO_3-N	2.7	mg	3/9
As	<10	µg	2/9
CCE	340	µg	1/8
Fe	40	µg	1/9

System 3 Substance	Median (per liter)		ER
NH_3-N	4.7	mg	9/9
TDS	582	mg	8/9
Cr	55	µg	5/9
NO_3-N	8.0	mg	4/9
Phenol	5	µg	3/9
CCE	400	µg	3/9
Ag	41	µg	2/9

TABLE 6-3 Examples of Nonspecific Measurements of Chemicals

Measure	Substance or Property Measured
Elemental parameters	Heavy metals, organic carbon, organic halogen
Collective organic parameters	Biochemical oxygen demand, chemical oxygen demand, total oxygen demand
Spectroscopic response, e.g., double bonds, aromatic organic compounds, or unshared electrons	UV, fluorescence
Analysis of chemical classes (functional group analysis)	Cholinesterase inhibitors (organophosphorous and carbamate pesticides), amines, carbohydrates, acids
Precursor analysis	Trihalomethane formation potential
Collective inorganic content	Conductivity

These nonspecific analyses, however, cannot replace analyses for specific inorganic or organic chemicals in wastewater renovation systems. Rather, they can serve as either surrogate measures for groups of specific chemicals that may be of health concern or, alternatively, as indications that (1) a process is not operating at specified levels, or (2) a predetermined limit has been exceeded, thus indicating a need for further analysis of the water for the presence of specific chemicals.

One well-studied example of a nonspecific parameter is TOC and its various subcategories. Dissolved organic carbon (DOC) is usually defined as the total value for a sample measurement after passage through a 0.45 μm filter. Purgeable organic carbon (POC) can be stripped from an aqueous solution at ambient temperatures in 10 minutes; nonvolatile TOC (NVTOC) is the organic carbon remaining after purging.

Known specific organic compounds of health concern have been estimated to be less than 10% of the organic content of drinking water (National Academy of Sciences, 1977). Thus, TOC measurements are not likely to be valuable surrogate measures for specific compounds. Nevertheless, surrogates are often useful in the design and routine monitoring of treatment systems.

Halogenated organic compounds are prevalent in water, and many chlorinated organic compounds are known to be of health concern (National Academy of Sciences, 1977). Thus, more useful nonspecific measurement, for significance to health, is that for organohalide (OX), which measures the quantity of chlorine, bromine, fluorine, and iodine in organic compounds. OX can be measured as total organohalide (TOX), purgeable organohalide (POX), and usually (by difference) nonpurgeable organohalide (NPOX). Measurements of these indices in water and wastewater have now advanced to the point that they can be performed by commercial instrumentation (Dressman et al., 1979; Takahaski, 1979) and they have been studied in wastewater reclamation systems (Jekel and Roberts, in press; McCarty et al., 1980).

Steigletz et al. (1976) reported that only about 10% by weight of the DOX in Rhine River water was identified by GC/MS analysis. It has been shown that both the DOX and the trihalomethanes (THM's) increase following the chlorination of natural waters (Kuhn and Fuchs, 1975; Oliver, 1978; Sander et al., 1977). However, THM's account for no more than 25% of the OX in drinking water (Yohe et al., 1980).

Organohalide and organic carbon are examples of nonspecific measurements that have a diverse utility for monitoring wastewater renovation systems. The closer such measures come to being indicators of, or surrogates for, chemicals of health significance, the greater the likelihood of their being incorporated into useful criteria for reused wastewater. In any event, they are useful for monitoring such systems because of their often greater practicality and simplicity, as compared to analyses for specific chemicals.

Specific Organic Analysis

The objective of efforts to analyze the organic content of water and wastewater samples comprehensively is to identify unequivocally as many organic components as possible. GC/MS is presently the analytical method of choice for this purpose. Popular awareness of this procedure has been increased markedly through a growing number of research reports and EPA's publication of standard procedures using GC/MS for analysis of "Priority Pollutants" (Environmental Protection Agency, 1979).

Experience suggests that the Priority Pollutant compounds comprise a small fraction of the TOC present in water and wastewater samples. Therefore, even good GC/MS techniques for detecting such pollutants might miss the vast majority of organic materials in the water, at least some of which may be at least as important toxicologically as the Priority Pollutants. Thus, concentration of efforts on the identification of Priority Pollutants alone may not be sufficient for judging the potential health risks associated with treated wastewater.

Standard identification procedures usually involve quadrupole-type mass spectrometers, and comparative identifications are based on computerized matching of mass spectra with those of reference compounds. This process is often insufficient to establish the identity of substances, i.e., simple matching of spectra can be misleading because of the occurrence of isomers, chemically similar compounds, and unresolved GC components.

Other problems exist with simple matching procedures. It is well documented (Budzikiewicz et al., 1967) that a given compound can often exhibit qualitatively different spectra on different mass spectrometers. Current data bases for identifications were compiled almost entirely using magnetic sector-type instruments, which can among themselves exhibit considerable variation in the spectra produced. Yet, since most current GC/MS work is done on quadrupole instruments, false positives and false negatives can occur in the matching process. Therefore, the results of simple matching procedures, even those done by computer, should be used cautiously.

These limitations are best avoided by comparing spectra of samples and reference compounds on the same model mass spectrometer and complementing these data with a compilation of retention index data of proposed structures on at least two different columns. This procedure constitutes a minimally acceptable set of criteria for structure identifications.

If, as is frequently the case, no match for a given mass spectrum can be found in the library or there is doubt in the matching process (e.g., if retention indices do not fit or spectra do not appear reasonable given the relative retention time), manual interpretation procedures must be applied, and further information will usually be required to postulate molecular structures. This additional information can be obtained by using current MS techniques. Successful identification depends heavily on the determination of molecular weight and molecular formula. Chemical ionization (CI) is a standard technique for obtaining and/or confirming molecular weight. Many commercial instruments can operate on both the electron impact (EI) and CI modes. Yet, this step is not currently part of any standard procedure.

The most important datum MS can provide is the elemental composition of an unknown compound. This technique requires the measurement of the mass of the molecular ion to the order of 10-15 ppm. This accuracy can be routinely achieved with some commercial instruments of the double-focusing type, but is well out of the range of most quadrupole instruments.

Because many of the compounds in wastewater samples have no (or doubtful) matches, it is important that these efforts be made at a suitable facility. Although provision of such data will not necessarily lead to 100% identification of all components, it will inevitably increase confidence in all structural assignments reached through computer matching and may make identification possible where matching fails.

With careful adoption of MS techniques, it is possible to investigate a large fraction of at least the chromatographable portion of the TOC in wastewater samples. An analysis that goes beyond Priority Pollutants, and perhaps well beyond the scope of many routine GC/MS analytical services, will be required.

Priority Pollutants

In 1979, EPA's Office of Waste Management published a report, <u>Water-Related Environmental Fate of 129 Priority Pollutants</u>. The pollutants listed in this report include the following classes: metals and inorganics; pesticides; polychlorinated biphenyls; halogenated aliphatic hydrocarbons; halogenated ethers; monocyclic aromatics; phthalate esters; polycyclic aromatic hydrocarbons; nitrosamines; and miscellaneous compounds (Table 6-4). In attempting to develop criteria and standards for inorganic and organic chemical constituents in the reuse of renovated wastewater for potable purposes, reedited and abbreviated priority listings for waterborne pollutants have been compiled (Andelman, in press; Environmental Protection Agency, 1977; Lichtenberg, in press; Suffet, in press). With no exceptions, all the listed chemical classes contain agents that have been shown to be carcinogenic in animal bioassays, and some classes contain chemicals that are carcinogenic in humans, according to epidemiological evidence. Moreover, all the classes contain mutagenic substances, a property that is closely correlated with carcinogenic activity.

Because of the human health hazards associated with these Priority Pollutant materials and the existence of relatively standard methods of analysis, it is reasonable to include these compounds in monitoring programs and preliminary testing studies at a frequency dictated by the level of positive findings.

"Sentinel Chemicals"

The term "sentinel chemicals" embodies the concept that a limited set of chemical compounds, which are typical representatives of major categories of substances believed to be hazardous to human health, can be used to provide a criterion against which the safety of drinking water may be determined. Sentinel chemicals ultimately selected for routine analysis could be based on its possession of the following characteristics:

- typical representative of its chemical structural class,
- frequent occurrence in finished drinking water,
- high stability against biodegradation,
- by-product of large-volume chemical manufacturing process(es);
- highly toxic (including carcinogenic or mutagenic) in some mammalian species or cell lines or produces chromosome abnormalities, or
- chlorination (or other halogenation) would probably increase the chemical's toxicity.

The chemicals constituting the sentinel set are likely to be unique to each reuse site and would be identified only after detailed preliminary testing. These sentinels would then be subjected to continued monitoring. This procedure provides for analyses of chemicals that may not appear on the list of Priority Pollutants, but that might contribute adverse health effects.

Infectious Agents and Their Indicators

It might seem that the best approach for determining the presence of infectious agents in water is to test for them directly. However, considering the different types of infectious agents that can be

TABLE 6-4 Groups of Priority Pollutants[a]

Metals
Antimony
Arsenic
Beryllium
Cadmium
Chromium
Copper
Lead
Mercury
Nickel
Selenium
Silver
Thallium
Zinc

Halogenated methanes (1 carbon)
Methyl bromide
Methyl chloride
Methylene chloride
   (dichloromethane)
Bromoform (tribromomethane)
Chloroform (trichloromethane)
Bromodichloromethane
Dibromochloromethane
Dichlorodifluoromethane
Trichlorofluoromethane
Carbon tetrachloride
   (tetrachloromethane)

Chlorinated (2 carbon)
Chloroethane (ethyl chloride
Chloroethylene (vinyl chloride)
1,2-Dichloroethane (ethylene
   dichloride)
1,1-Dichloroethane
1,2-trans-Dichloroethylene
1,1-Dichloroethylene (vinylidene
   chloride)
1,1,2-Trichloroethane
1,1,1-Trichloroethane (methyl
   chloroform)
Trichloroethylene
Tetrachloroethylene
1,1,2,2-Tetrachloroethane
Hexachloroethane

Chlorinated (3 carbon)
1,2-Dichloropropane
1,3-Dichloropropylene

Chlorinated (4 carbon)
Hexachlorobutadiene

Chlorinated (5 carbon)
Hexachlorocyclopentadiene

Chloroalkyl ethers
Bis(chloromethyl)ether
Bis(2-chloroethyl)ether
Bis(2-chloroisopropyl)ether
2-Chloroethylvinyl ether
Bis(2-chloroethoxy)methane

Pesticides
Aldrin
Dieldrin
Chlordane
α-Endosulfan
Endrin
Endrin aldehyde
Heptachlor
Heptachlor epoxide
α-BHC
β-BHC
γ-BHC (lindane)
δ-BHC
4,4'-DDT
4,4'-DDE ($\underline{p},\underline{p}$'-DDX)
4,4'-DDD ($\underline{p},\underline{p}$'-TDE)
Toxaphene

Nitrosamines
$\underline{N}$-Nitrosodimethylamine
$\underline{N}$-Nitrosodiphenylamine
$\underline{N}$-Nitrosodi-$\underline{n}$-propylamine

Miscellaneous
Acrolein
Acrylonitrile
Isophorone
Cyanide

Aromatics
Benzene
Toluene
Ethyl benzene

Table 6-4 Continued

Polyaromatics
Naphthalene
Acenaphthene
Acenaphthylene
Anthracene
Benzo[a]anthracene(1,2-
 benzanthracene)
Benzo[a]pyrene(3,4-
 benzopyrene
3,4-Benzofluoranthene
Benzo(k)fluoranthene(11,12-
 benzofluoranthene)
Benzo(ghi)perylene(1,12-
 benzoperylene)
Chrysene
Dibenzo(a,h)anthracene(1,2,5,6-
 dibenzanthracene)
Fluorene
Fluoranthene
Indeno(1,2,3-cd)pyrene(2,3-o-
 phenylene pyrene)
Phenanthrene
Pyrene

Chloroaromatics
Chlorobenzene
o-Dichlorobenzene
p-Dichlorobenzene
m-Dichlorobenzene
1,2,4-Trichlorobenzene
Hexachlorobenzene

Chlorinated polyaromatic
2-Chloronaphthalene

Polychlorinated biphenyls
Seven listed

Phthalate esters
Bis(2-ethylhexyl)phthalate
Butylbenzyl phthalate
Di-n-butyl phthalate
Di-n-octyl phthalate
Diethyl phthalate
Dimethyl phthalate

Nitroaromatics
Nitrobenzene
2,4-Dinitrotoluene
2,6-Dinitrotoluene

Benzidines
Benzidine
3,3'-Dichlorobenzidine
1,2-Diphenylhydrazine

Phenols
Phenol
2,4-Dimethylphenol

Nitrophenols
2-Nitrophenol
4-Nitrophenol
2,4-Dinitrophenol
4,6-Dinitro-o-cresol

Chlorophenols
2-Chlorophenol
4-Chloro-m-cresol
2,4-Dichlorophenol
2,4,6-Trichlorophenol
Pentachlorophenol
TCDD(2,3,7,8-tetrachlorodibenzo-
 p-dioxin)

Haloaryl ethers
4-Chlorophenylphenyl ether
4-Bromophenylphenyl ether

[a]From Environmental Protection Agency, 1979.

present in a raw domestic wastewater, treated effluent, receiving water, or the public water supply, this approach would require a vast number of tests, some of which involve complex, time-consuming, and often insensitive procedures. Furthermore, the density of different infectious agents varies in different waters, including wastewater. This variability, which makes the detection of infectious agents difficult and somewhat unreliable, is a function of the number of intestinal infections that can occur at different times in the contributing warm-blooded population. In recognition of these constraints, it has been common practice to use microbial indicators or surrogates to indicate fecal contamination of water. Because the source of most infectious agents is the feces of humans and other animals, the focus has been on the use of organisms (indicators) that occur naturally in the feces of warm-blooded animals. Such indicators suggest the presence or absence of fecal contamination in water and, at the same time, imply the presence or absence of infectious agents.

To be a satisfactory indicator, an organism or group of organisms should be present in the feces of warm-blooded animals at a density greater than that for any infectious agent so as to be easily and unambiguously detectable. There should be a positive correlation between the indicator and fecal contamination. There should also be some correlation between the response of the indicator and the various infectious agents to different environmental conditions and treatment processes (e.g., disinfection).

Although there are other requirements for an acceptable indicator, these are perhaps the most important. Over the years, a number of different microbial indicators have been proposed. Some used more commonly than others are discussed below.

Total Coliforms

Since it was proposed some 70 years ago, the most widely used microbial indicator group for determining the presence and intensity of fecal contamination in water has been the coliform bacteria. The density of total coliforms in raw domestic wastewater may be as high as 10^9/100 ml, but in the United States the density most commonly ranges from 10^6 to 10^8/100 ml (Geldreich, 1978; World Health Organization, 1975).

Over the years, the total coliform test has become an accepted indicator of fecal contamination, and there is ample evidence to justify this acceptance (National Academy of Sciences, 1980). The occurrence of any coliform bacteria in water should signal possible fecal contamination and the likelihood that infectious agents may be present.

Because of its general acceptance, even with respect to monitoring the adequacy of water treatment, the total coliform indicator system is used as the basis for the EPA's National Primary Drinking Water Regulations for microbial agents. In practice, the absence of total coliforms in a potable water supply has been considered adequate evidence of a microbiologically safe supply; however, there have been instances where the total coliform test has not signaled the presence of waterborne Salmonella and Giardia

(McFeters et al., 1978). Furthermore, the coliform group of bacteria is less resistant to chlorination, the most widely used means of water disinfection, than are some other infectious agents, such as protozoan cysts and enteric viruses.

Fecal Coliforms

The use of the total coliform as an indicator of fecal contamination has been criticized because some strains of coliforms are widely distributed throughout the environment. These strains, therefore, are not directly associated with fecal matter, but may be present in wastewater. To overcome this disadvantage, the fecal coliform test has been used as perhaps a more exact measure of fecal contamination of water. Fecal coliforms constitute more than 90% of the total coliforms normally found in the feces of warm-blooded animals (Geldreich, 1978). The density of fecal coliforms in raw domestic wastewater ranges from 10^5 to 10^6/100 ml. Although fecal coliforms may be more directly related to fecal contamination and their measure may have certain other advantages over the total coliform indicator system (e.g., less regrowth in contaminated water), there remains the need to correlate the occurrence of fecal coliforms in water with the presence of waterborne infectious agents. The total coliform indicator system has a demonstrated value in assessing the occurrence of infectious agents in water, but its results may be conservative because it offers a margin of safety not associated with the fecal coliform test (Geldreich, 1978).

Fecal Streptococci

The fecal streptococci group of bacteria has been used as a measure of fecal contamination for approximately 50 years. Raw domestic wastewater may contain up to 10^6 fecal streptococci/100 ml (Geldreich, 1978). As a group, the fecal streptococci include strains having variable survival rates and others having little significance as indicators of fecal contamination because of their occurrence and reproduction outside of the intestinal tract of warm-blooded animals. Fecal streptococci can exist for extended periods of time in certain waters.

Clostridium perfringens

This anaerobic bacterium is commonly found in the feces of warm-blooded animals, and its use as an indicator of fecal contamination has been rigorously supported in certain areas of Europe. However, the organism is also regarded as being ubiquitous in nature (McFeters et al., 1978). Because it forms spores, Clostridium perfringens persists longer in water than do non-spore-forming bacteria such as coliforms.

Standard Plate Count

This technique, as currently used, measures the number of heterotrophic bacteria in water. These microorganisms can grow at 35°C. The standard plate can be as high as $10^{10}/100$ ml in raw wastewater and appears to provide additional valuable information relative to the presence of infectious agents in water.

Other Indicators

A number of other indicators have been proposed, and some of them have been used to a limited extent. For the most part, there is not enough information to determine their suitability or utility as indicators of fecal contamination. The use of Pseudomonas spp. as an indicator group has been proposed, but they are ubiquitous in nature and actively reproduce in potable water and even in distilled water; thus, this indicator has not generated much attention (McFeters et al., 1978). Furthermore, because Pseudomonas can persist for an extended period in water, it has been observed in the absence of coliform bacteria.

Lactobacillus, Bifidobacterium, and Bacteroides have also been proposed as indicators (Geldreich, 1978). These non-spore-forming anaerobic bacteria occur generally at higher densities in feces than do aerobic bacteria and frequently outnumber the coliforms. Unfortunately, not all humans shed a detectable number of Lactobacillus and Bifidobacterium in fecal matter. This observation, plus their poor survival in the aquatic environment, limit their utility as indicators of fecal contamination. Enumeration methodology is also a problem with these anaerobic bacteria, particularly in the case of the Bacteroides, which are obligate anaerobes.

The use of species of Aeromonas as an indicator of fecal contamination of water has also been suggested. Members of this genus of bacteria are widely distributed in nature. Some of them are infectious to aquatic animals (McFeters et al., 1978) and many are capable of long survival in the aquatic environment. For these and other reasons, Aeromonas has not generated much attention as an indicator.

It has also been proposed recently that acid-fast bacteria might be attractive as an indicator of disinfection efficacy because their sensitivity to chlorine is similar to that of several enteric viruses that have been observed to have a resistance greater than that of the coliforms (Engelbrecht et al., 1979). Acid-fast bacteria appear to be common inhabitants of raw domestic wastewater.

The bacteriophage, particularly the coliphage, has also been suggested as a suitable indicator of enteric viruses and, perhaps, other infectious agents in water. Depending on the host culture, raw domestic wastewater may contain up to 10^6 coliphage/100 ml (Scarpino, 1978). Although limited data indicate that there may be a correlation between the coliphage and enteric viruses in water, other data indicate that the coliphage is an inadequate indicator. Noncoliphage organisms, such as Serratia marcescens and the staphylococcal bacteriophage, have also been proposed as indicators of enteric viruses

(Scarpino, 1978). The bacteriophage may be particularly useful in evaluating removal or inactivation of viruses by wastewater and water treatment processes. It appears, however, that much remains to be learned about the bacteriophage in water, including wastewater.

Finally, the measurement of fecal sterols, adenosine triphosphate (ATP), and bacterial endotoxins in water have been studied as rapid indicators of fecal contamination or as providing possible supplementary information. For example, the Limulus endotoxin assay has been investigated for correlation with conventional bacterial indicator tests (Jorgensen et al., 1979). When this assay has been used on environmental samples, the significance and interpretation of the results have been questionable. Thus, the utility of this test needs additional evaluation.

Ideally, a microbiological indicator should be selected that, when present in concentrations above an established limit, implies that the ingestion of the water poses an unacceptable risk of infecting its consumers. The results of the tests should be available before final distribution (real-time indicator); if the limit is exceeded, the water should receive additional treatment or another source of drinking water should be obtained.

MONITORING STRATEGIES

In view of the many uncertainties regarding chemical and biological composition of wastewater influents to water renovation plants, both extensive preliminary and routine analyses should be performed. The cost of this activity for reuse will be considerably greater than for ordinary water supply development. Special emphasis should be placed on analytical efforts before a renovation facility is actually operated. Benefits from such investments will include fewer uncertainties about the variability in quality of the wastewater to be renovated. Several months of sampling may be required to determine the day-to-day variation in water quality to be expected at each site and to identify major industrial and urban contributors to the wastewater flow streams.

Monitoring Frequencies

Determining the frequency for monitoring is complicated. Among the factors that must be considered are the diurnal variability, spatial variability, seasonal variability, and desirable detection levels. Consideration of these factors implies that site-specific monitoring programs need to be implemented. In-depth preliminary studies should be conducted to determine ambient conditions and their stability. Subsequent monitoring is then needed to detect shifts in the quality of input. The treatment process (at critical points) and the effluent should be monitored in accordance with predetermined detection levels. After preliminary studies, the monitoring frequency might reasonably shift primarily to the renovation plant operation itself, i.e., influent, major treatment units, and effluent, along with any prerenovation plant system inputs determined in the preliminary study to be particularly heavy industrial contributors.

Ideally, detection limits for chemical parameters should be determined by maximum allowable concentrations established for the protection of public health. The limits can then be used to determine the necessary sample volume. Information gathered in the preliminary study and monitored at the influent can be used to specify frequency of sampling. Biological parameters will be similarly monitored, but the amount of sampling should be determined by a different statistical model. Unless preliminary data indicate otherwise, a log-normal model seems appropriate for the chemical concentrations and the Poisson model for the biological parameters. Continual checking using "goodness-of-fit" procedures should be performed to detect model inadequacies.

Sampling for and monitoring of viruses pose a special problem. Unlike analyses for specific chemicals, current methods for recovery and detection of viruses from large volumes of water are subject to a high degree of uncertainty.

In sampling for viruses, let λ = virus density (number per unit volume), p = proportion counted (recovery rate), and V = volume. The expected number of viruses in volume V is $p\lambda V$. The probability of observing no viruses in volume V is $\exp[-p\lambda V]$. The upper 95% confidence limit on virus density is given by

$$\lambda_u = -\ln 0.05/pV = 3/pV.$$

Thus, if the recovery rate of viruses is 25% and if no viruses are discovered in 1,000 gallons of water, the upper 95% confidence limit is $3/0.25 \times 1,000 = 0.012$ or 1 virus per 83 gallons.

Additional mathematical details on sampling and monitoring are presented in Appendix B.

REFERENCES

Andelman, J.B. In press. Inorganic chemicals in reuse systems. Criteria and Standards for Potable Reuse and Feasible Alternatives, Appendix II. Airlie House Workshop, July 29-31, 1980. Environmental Protection Agency, Washington, D.C.

Budzikiewicz, H., C. Djerassi, and D. Williams. 1967. P. 9 in Mass Spectrometry of Organic Compounds. Holden-Day, Inc., San Francisco, Calif.

Dressman, R.C., B.A. Najar, and R. Redzikowski. 1979. The analysis of organohalides (OX) in water as a group parameter. P. 69 in 7th Annual AWWA Water Quality Technology Conference, Dec. 9-12, Philadelphia. American Water Works Association, Denver, Colo.

Engelbrecht, S.R., C.N. Hass, J.A. Shular, D.L. Dunn, D. Roy, A. Lalchandani, B.F. Severin, S. Farooq, and R.H. Taylor. 1979. Acid-Fast Bacteria and Yeasts as Indicators of Disinfection Efficiency. EPA-600/2-79-091. Municipal Environmental Research Laboratory, Office of Research and Development, U.S. Environmental Protection Agency, Cincinnati, Ohio. 143 pp.

Englande A.J, Jr., and R.S. Reimers III. 1979. Wastewater reuse--
Persistence of chemical pollutants. Pp. 1368-1389 in Proceedings,
Water Reuse Symposium. American Water Works Association Research
Foundation, Denver, Colo.
Environmental Protection Agency. 1977. National Organic
Monitoring Survey--General Review of Results and Methodology, Phase
I-III. Office of Water Supply, Technical Support Division,
Environmental Protection Agency, Washington, D.C.
Environmental Protection Agency. 1979. Water-Related Environmental
Fate of 129 Priority Pollutants. Office of Waste Management,
Environmental Protection Agency, Washington, D.C.
Geldreich, E.E. 1978. Bacterial populations and indicator concepts
in feces, sewage, stormwater and solid wastes. In G. Berg, ed.
Indicators of Viruses in Water and Food. Ann Arbor Science, Ann
Arbor, Mich.
Hart, O.O. 1978. Operational guidelines. Chapter 13 in Manual
for Water Renovation and Reclamation. Technical Guide K42.
National Institute for Water Research, Council for Scientific and
Industrial Research, Pretoria, Republic of South Africa.
Jekel, M., and P.V. Roberts. In press. Total organic halogen
measurements for the characterization of reclaimed waters:
Occurrence, formation, and removal. Environ. Sci. Technol.
Jorgensen, H.H., J.C. Lee, G.A. Alexander, and H.W. Wolf. 1979.
Comparison of *Limulus* assay, standard plate count, and total
coliform count for microbiological assessment of renovated
wastewater. Appl. Environ. Microbiol. 37:928-931.
Kuhn, W., and F. Fuchs. 1975. Investigations on the significance
of organic chlorine compounds and their absorbability. Von Wasser
45:217-232
Lichtenberg, J.J. In press. Analytical methods that are needed
for organic chemicals of concern to health. Criteria and Standards
for Potable Reuse and Feasible Alternatives, Appendix I. Airlie
House Workshop, July 29-31, 1980. Environmental Protection Agency,
Washington, D.C.
McCarty, P.L., J. Kissel, T. Everhart, R. Cooper, and C. Leong.
1980. Mutagenic activity and chemical characterization for the
Palo Alto Wastewater Reclamation and Groundwater Injection Facility.
Tech. Rep. No. 250. Health Effects Research Laboratory, U.S.
Environmental Protection Agency, Cincinnati, Ohio. 65 pp.
McFeters, G.A., J.E. Schillinger, and D.G. Stuart. 1978. Alter-
native indicators of water contamination and some physiological
characteristics of heterotrophic bacteria in water. Pp. 37-48 in
C.W. Hendricks, ed. Evaluation of the Microbiology Standards for
Drinking Water. EPA 570/9-78-00C. Office of Drinking Water,
Environmental Protection Agency, Washington, D.C.
National Academy of Sciences. 1977. Drinking Water and Health.
Safe Drinking Water Committee, Advisory Center on Toxicology,
Assembly of Life Sciences, National Research Council. National
Academy of Sciences, Washington, D.C. 933 pp.
National Academy of Sciences. 1980. Drinking Water and Health,
Vol. 2. Safe Drinking Water Committee, Board on Toxicology and
Environmenal Health Hazards, Assembly of Life Sciences, National
Research Council. National Academy Press, Washington, D.C. 393 pp.

Oliver, B.G. 1978. Chlorinated non-volatile organics produced by the reaction of chlorine with humic material. Can. Res. 11:21.

Sander, R., W. Kuhn, and H. Sontheimer. 1977. Study of the reaction of chlorine with humic substances. Z. Wasser Abwasser Forsch. 10(5):155-160.

Scarpino, P.V. 1978. Bacteriophage indicators. Pp. 201-227 in G. Berg, ed. Indicators of Viruses in Water and Food. Ann Arbor Science, Ann Arbor, Mich.

Steiglitz, L., W. Roth, W. Kuhn, and W. Leger. 1976. Behavior of organic halogen compounds in the purification of drinking water. Von Wasser 47:347-377

Suffet, I.H. In press. Issue paper--Chemistry. Criteria and Standards for Potable Reuse and Feasible Alternatives. Airlie House Workshop, July 29-31, 1980. Environmental Protection Agency, Washington, D.C.

Takahashi, Y. 1979. Analysis techniques for organic carbon and organic halogen. Proceedings of EPA/NATO-CCMS Conference on Adsorption Techniques, Reston, Va.

World Health Organization. 1975. Health Effects Relating to Direct and Indirect Reuse of Wastewater for Human Consumption. Report of an International Working Meeting, held at Amsterdam, The Netherlands, Jan. 13-16, 1975. World Health Organization Tech. Pap. No. 7. The Hague, The Netherlands. 164 pp.

Yohe, T.L., I.H. Suffet, and R.J. Grochowski. 1980. Development of a teflon helix continuous liquid-liquid extraction apparatus and its application for the analysis of organic pollutants in drinking water. Pp. 47-67 in C.E. Van Hall, ed. Measurement of Organic Pollutants in Water and Wastewater. ASTM STP 686. American Society for Testing and Materials, Washington, D.C.

7
Assessment and Criteria for Potable Water Reuse

The Panel on Quality Criteria for Water Reuse strongly endorses the generally accepted concept that drinking water should be obtained from the best quality source available. Because the costs of wastewater treatment for potable use are high, it is anticipated that reuse would be contemplated only in the few locations where alternatives are nonexistent or even more costly. In such events, criteria to judge the relative safety of the treated wastewater for human consumption are needed. Although it was for such a setting that the criteria suggested in this chapter were developed, it is anticipated that they may also be useful for locations faced with significant indirect potable reuse.

When the risk of treated wastewater is to be evaluated, an experimental facility is required to evaluate process performance and quality of treated wastewater under normal conditions and with the variations expected from operation of a full-scale system. To improve the possibility of the wastewater's meeting acceptable criteria, it should be a domestic wastewater selected from a source containing as little industrial and agricultural waste and urban runoff as possible. The treatment systems selected should use a series of processes that offer some redundancy in their capability for removing the chemicals present. Operation of the plant is best with a constant flow, and added safety is generally provided by allowing production to be stopped in response to mechanical or influent wastewater quality problems. Good monitoring and control features are also necessary to provide adequate performance reliability.

Finally, provision of storage for the treated water would allow time for further natural decay processes to occur and to permit routine quality testing for making decisions about acceptability of the effluent for use by humans. A reuse treatment system must be monitored and controlled with full recognition of the potential hazards to the public that could result from operational failures. With these limitations and considerations in mind, the panel developed the following criteria for evaluating data describing the risks presented by consumption of treated wastewater relative to those from consumption of conventionally treated tap water.

RECOMMENDED ANALYSES AND TESTS FOR QUALITY EVALUATION

The panel's approach to establishing quality criteria for water reuse relies heavily on the belief that the relative degree of potential hazard to human health from treated wastewater can be estimated by comparative biological testing and toxicological testing of concentrates from treated wastewater and conventional water supplies. The scientific considerations connected with this estimation are the subject of the balance of this report.

The thrust of the panel's effort was directed toward estimating relative risk from the consumption of treated wastewater; the criteria, therefore, involve comparative toxicological testing procedures and their interpretation. The panel did not address other criteria, such as determining the degree of acceptability associated with any relative potential hazard, e.g., economic considerations, availability of alternatives, and selected indices of public preference.

There has been little experience with or understanding of the potential or actual adverse health effects associated with the use of treated wastewater for human consumption either through drinking or its use in food processing. Therefore, the emphasis of this panel's recommendations is that there be extensive analytical monitoring and toxicity studies at the pilot-plant stage of a planned reuse system. At this time, such studies cannot be considered routine and certainly are not standardized or specified in advance as a rigid, predetermined testing scheme. Indeed, in many ways the tests may have to be considered as research efforts, the results of which will be used not only to provide a basis to assess the desirability to proceed with actual reuse but also to provide the information required to develop a monitoring scheme for reused water should it become the option of choice. However, it is also reasonable to anticipate that, with the successful development and operation of such systems, more standardized testing protocols and evaluation methodologies will be generated in the future. Thus, the following sections propose testing schemes for pre-reuse (pilot) studies to provide a data base to which would be applied the criteria set forth in this chapter. The panel recommends that three types of data be initially monitored: drinking water standards, individual chemicals and microbiological organisms, and the chemical fingerprint of the concentrated mixtures.

Drinking Water Standards

The analytical and monitoring requirements of the Safe Drinking Water Act (PL 93-523) apply to any reuse effort where the effluent from the renovation facility is directly connected to the water supply distribution system. In cases resulting in indirect reuse, these requirements must be met by the intake treatment process. In either case, it is assumed that the primary and secondary drinking water standards (Environmental Protection Agency, 1975, 1977) will be met along with the monitoring and quality assurance requirements stipulated in the

Safe Drinking Water Act. Furthermore, it is assumed that there will be compliance with all additions and revisions to these standards.

Chemical Analysis

It is extremely important that every effort be expended to establish the inorganic and organic composition of the influent and effluent of the pilot or treatment facility. This analysis will involve surveys of all major domestic and industrial inputs to the wastewater collection system, and of the influent to the treatment facility, to establish source and time variations. Efforts should be made to characterize the removal efficiencies of major unit operations in the treatment train with regard to trace metals and specific organic contaminants. Concentration ranges for toxic trace elements such as arsenic, cadmium, chromium, copper, lead, mercury, selenium, and silver must be known or predicted.

Similarly, the composition and concentrations of major organic contaminants must be established. The effort will unquestionably involve the application of relatively sophisticated gas chromatographic/mass spectrometric (GC/MS) techniques. Useful guidelines for organic analyses have been suggested by McCarty et al. (1980). All organic constituents present in detectable concentrations (at least 1 µg/liter) should be identified; special attention should be paid to aromatic hydrocarbons, synthetic chlorinated compounds, chlorination products, natural products, phthalate esters, and miscellaneous compounds identified in similar large-scale treatment projects. In addition to screening and analyzing detectable organic constituents, it is also recommended that a wide range of general and specific chemical parameters be targeted for analysis. Along with the potentially toxic inorganic substances mentioned above, monitoring should be conducted to include the Environmental Protection Agency's (EPA) Priority Pollutants for which health criteria documents have been developed, as well as other specific chemicals likely to be present and about which there is information on adverse health effects. Standardized methodologies have been developed for identification and measurement of the Priority Pollutants.

Although the utility of surrogate chemical monitoring is subject to question for the reasons discussed in earlier chapters, preliminary and pilot-scale testing should involve analysis of surrogates (e.g., total organic carbon and total organic halogens) to uncover relationships, if any, that may have predictive value for changes in the effluent or its concentrates at the reuse site under development. These tests are relatively inexpensive and rapid; they provide a data base for correlations that may become evident only in the future.

A special analytical problem is presented by the need to develop concentrates for toxicological testing. It is necessary to determine the inorganic and organic composition (chemical "fingerprint") of the concentrate. This procedure involves extremely careful and extensive GC/MS analytical work, which, when compared with the inorganic and surrogate chemical data, will provide the only chemical indication of artifact formation or constituent loss as a result of the concentration process.

Microbiological Analysis

Acceptance of drinking water for human consumption requires the development and implementation of limits based on the protection of public health from waterborne microbiological diseases. The federal standards for conventionally treated water are listed in the Interim Primary Drinking Standards Water (Environmental Protection Agency, 1975). For water reclaimed for potable use, the guidelines can be based on specific infectious agents, appropriate indicator system(s), and/or required treatment. A guideline based on each specific infectious agent is likely to be impractical because of the different types and densities that might be present, which in turn would necessitate a large number of different tests. Furthermore, the current detection and enumeration methods for viruses, protozoa, helminths, and some bacteria are inadequate. The use of an indicator organism is more practical and is currently being used for coliform organisms; in fact, the maximum contaminant level (MCL) for the microbiological quality of drinking water is currently based solely on coliform bacteria (Environmental Protection Agency, 1975). In view of the available data base, the coliform test is adequate as an indicator of the infectious enteric bacteria, and perhaps other infectious agents, when a high quality raw water source is used for a water supply. However, for water reclaimed for potable use, the current coliform test does not assess viral quality of the water, and its adequacy for indicating the presence or absence of some pathogenic bacteria, infectious protozoa, and helminths is uncertain. Studies on the reliability of the coliform bacteria as an acceptable indicator are needed, particularly with respect to assessing the microbiological quality of water reclaimed for potable use. Until the adequacy of the coliform test as an indicator of the above pathogens is established, specific tests for viruses, infectious protozoa, helminths, and _Salmonella_ should be conducted on any reclaimed water intended for potable use. Furthermore, other indicator systems, such as coliphage, acid-fast bacteria, and the standard plate count, should be studied as potential indicators of the different infectious agents that might be found in reused water. In this respect, no single indicator system may be adequate; thus, several indicators must be used for reliable assessment of the microbiological quality of a reclaimed water.

Toxicity Testing of Concentrated Mixtures

The panel proposes that reused water concentrates of organic substances be toxicologically evaluated on a comparative basis with conventional water concentrates from the same geographical location. Testing should progress through three phases of evaluation. In Phase 1, short-term _in vitro_ and _in vivo_ assays have been identified to assess mutagenic, carcinogenic, and teratogenic potential; target organ toxicity; and clastogenic activity. Phases 2 and 3 should evaluate subchronic and chronic effects that would not become evident within the time frame of the Phase 1 assays. Progression of testing into Phases 2 and 3 should be dependent on the nature of the results in Phase 1 or a determination of the relative risk from reused water compared to a conventional water. The Phase 2 study is a 90-day subchronic evaluation of possible cumulative adverse effects such as target organ toxicity, aberrant behavioral and physiological functions, and histological evidence of tissue alteration. Adverse reproductive effects should also be evaluated in Phase 2. Phase 3 is a combination chronic toxicity and carcinogenicity study.

The tests identified for the Phase 1 evaluation can be performed in a relatively short time, permit evaluation of multiple samples from a reused water purification process, and characterize some forms of irreversible toxicity, if any, that are generally of the greatest concern. Table 5-1 (in Chapter 5) outlines the several phases of toxicological evaluation.

CRITERIA FOR WATER REUSE EVALUATION
SELECTION OF CONVENTIONAL WATERS FOR COMPARISON WITH REUSED WATER

There appears to be no scientific or societal consensus as to what constitutes an "ideal" potable water. Potability is determined by acceptability of taste and odor and the presumed absence of unacceptable adverse health effects. In the absence of an absolute, ideal water standard, the performance of a wastewater treatment facility to produce potable water should be judged in comparison with conventional drinking waters. The philosophy behind the Interim Primary Drinking Water Regulations requires that water intended for human consumption should be taken from the highest quality source that is economically feasible. Accordingly, in assessing the adequacy of water being considered for potable reuse, comparision should be made with the highest quality of water that can be obtained from that locality even though that source may not be in use.

The quality of conventional water sources varies from location to location. With increased industrialization and urbanization, it seems reasonable to assume that the quality of the sources of some conventional potable water supplies will gradually change and possibly deteriorate. Because the panel recommends that the quality of reused water should be determined by comparison to conventional waters, it would be useful to establish a registry of the composition of conventional water in all areas where potable reuse is contemplated or in practice. This registry would represent a compilation of the known chemical and microbiological compositions of registered conventional waters, determined by standard testing procedures. Comparative toxicity testing (as described in the report) should also be performed periodically for waters listed in the registry to serve as a baseline for future comparative testing.

The registry would have two important benefits. First, it could be used to provide a wider-than-local comparative perspective by making it possible to compare the toxicological properties of treated wastewater at any location with those from any other source selected from the registry. Second, the existence of such a registry would permit the preliminary evaluation of a proposed treated wastewater supply with other local optional sources, if any exist.

Assessment of Results of Comparative Studies

An objective of pilot-plant studies is to evaluate the performance and reliability of a given treatment system or of alternative treatment systems for the reclamation of a given source of wastewater. The operation under a selected mode should be of sufficiently long duration to encompass seasonal changes in wastewater quality. Data for both chemical and microbiological analyses and toxicity testing should be representative of at least 1 full year of operation. Sufficient data must be collected to provide an adequate data base for statistical analysis. The quality of the reclaimed water should then be compared with that of conventional sources available to the community in a comparative risk assessment and, wherever appropriate, with other sources such as those included in the suggested registry described previously. Two phases of comparison are proposed: one in relation to results of analyses of specific constituents or group parameters and the other as related to results of toxicity testing of concentrates.

Microbiological Criteria

Some criteria have to be applied to ensure the microbiological acceptability of a water supply. Heretofore, it has been common practice to develop a public water supply from the highest quality water source available; for the most part, this practice has tended to minimize the risk of transmitting infectious disease. Currently, for a water to be potable, it must meet a maximum contaminant level (MCL) for coliform bacteria, regardless of whether or not the source water has been indirectly affected by wastewater discharges. Even with source waters of reasonably good quality, there have been instances where the existing MCL has not been 100% effective in protecting the public against waterborne disease. Where a source water of poor microbiological quality is to be used, such as with potable reuse, more rigorous quality criteria are required. It is generally agreed that there should be no detectable pathogenic agents in water intended for human consumption.

Guidelines need to be related to particular sites in the overall system, and monitoring of these sites is necessary to ensure that the guidelines are maintained. The obvious sites to monitor for a direct potable water reuse system are (1) the wastewater treatment plant influent and effluent, (2) treated water, and (3) the water distribution system. The need for real-time monitoring of the treated water before it is pumped into the water distribution system should be evaluated in detail. Monitoring of the water distribution system should cover all areas of the system and should be sufficiently frequent to provide a reasonable probability of determining the presence of infectious agents.

In the case of potable reuse, there does not appear to be any major advantage gained by substituting the fecal coliform test for the total coliform test as an indicator system. That total coliforms exist in greater number than do fecal coliforms in wastewater provides an additional margin of safety. Because raw water sources directly affected by wastewater clearly possess a greater health risk than do conventional waters, the current MCL for potable water (based on the total coliform test) is not adequate for reclaimed water, because the surrogates do not cover all forms of microbial agents. With potable reuse, the ideal would be a complete absence of total coliform bacteria in samples of at least 1 liter; however, this may not be possible.

Coliform bacteria may be an adequate indicator for *Salmonella* and *Shigella* but perhaps not for all bacteria and other infectious agents that might be present in domestic wastewater. The possible presence of other infectious agents must be considered with potable reuse and, in this case, perhaps a new indicator system(s) identified. The standard plate count, for example, might be useful as a supplement to the total coliform test when monitoring the general bacterial quality of the product water and assessing of disinfection efficiency.

In considering a virus standard for reclaimed drinking water, a World Health Organization (1979) scientific group concluded that "no viruses should be detectable in samples of between 100 and 1,000 liters." Because there remains a need for a better understanding of the reliability, limit of detection, and precision of the current methods used for virus enumeration, there does not seem to be any way at present to establish a meaningful guideline for viruses. If such a virus guideline could be established, based on current detection methods, there would still be problems in applying the methods to routine, real-time monitoring. That is, when tested for enteric viruses, a minimum 2-week period is required before the results from a sample of water can be reported. It would seem reasonable, therefore, to identify a suitable indicator system for assessing the viral quality of a potable water.

Currently, there is no standard procedure for the detection and enumeration of intestinal protozoan cysts or helminths in water; consequently, it is not now feasible to consider establishing a guideline for their presence. A reliable procedure to detect them directly in water is needed. The relationship between coliform bacteria and the presence or absence of infectious parasites in water is also currently unknown.

It seems desirable to consider the development of MCL values, together with the establishment of criteria regarding the removal and/or inactivation of infectious agents and indicators of infectious agents achieved through water treatment. Included in this approach is the need to have appropriate indicator systems that rapidly, reliably, and (with an acceptable degree of sensitivity) continuously monitor the microbiological quality of the product water.

Until such time that adequate detection methods and more information are available regarding various indicator systems, it is recommended that, based on the total coliform test, a practical limit of less than 1 coliform/1,000 ml at least 90% of the time and less than 1/100 ml at least 98% of the time be established for comparison with conventional waters; sampling frequency should be at least daily. For viruses, there should be less than 1/1,000 liter. Testing for viruses, as well as for protozoan cysts and helminth ova, should be performed weekly using the best available detection methods. It is further recommended that, based on daily sampling, the standard plate count bacteria not exceed 100/ml. These recommendations are judged to be reasonable in light of existing information. Although not part of the specific recommendations, it is suggested that information on the applicability and suitability of other indicator systems (e.g., coliphage and acid-fast bacteria) be obtained at the same time as the above recommended testing.

Constituent Chemical Analyses

The comparison of constituent analyses for the reclaimed and conventional waters includes first an evaluation to see if both meet existing local water quality standards such as the EPA primary and secondary drinking water regulations. If the reclaimed water does not meet such standards, then it should be considered as carrying a greater risk for drinking water and for use in food processing and, at a minimum, should be subjected to further testing.

Next, a comparison of the microbial constituents and indicators, including viral analysis, should be performed. Comparisons of constituent analyses should follow for a number of other (unregulated) specific inorganic and organic compounds for which there currently are no standards, but whose concentrations are readily measured and which are believed to be potentially hazardous. The presence of such chemicals, e.g., EPA Priority Pollutants or sentinel chemicals (see Chapter 6), would constitute some degree of potential health risk, the magnitude of which would be influenced by the nature (i.e., concentration and duration) of exposure.

If the results of pilot-plant testing indicate that reclaimed water meets current drinking water standards, and if all measures of biological contamination and other individual constituents indicate that the quality of reclaimed water is at least as good as that of water representative of "conventional" sources, then it can be concluded that no greater health risk has been demonstrated for the reclaimed water, <u>with respect to these measured parameters only</u>, than for the water from the conventional source. One basis for such a comparison is provided in the Safe Drinking Water Committee's reports entitled <u>Drinking Water and Health</u> (National Academy of Sciences, 1977, 1980, 1982). Such a comparison would indicate if the reclaimed water presents a larger, smaller, or essentially the same health risk as does the conventional water. If the comparison of results from constituent analyses indicates that the reclaimed water compares favorably in quality with that from conventional sources, then the next step would be to compare the results from toxicity testing of concentrates.

Since it is unlikely that such a favorable comparison would exist for all measures of water quality, further evaluation would be required for the constituents found in higher concentrations in the treated wastewater. One alternative then would be to alter the pilot treatment system or the wastewater source to improve the quality of the reclaimed water in an attempt to lower the levels of the constituents in question. The conventional source may impose more risk with respect to some constituents, and the reclaimed source with respect to others. An evaluation of existing toxicological data for

each of the constituents of concern would be necessary in making such
a comparative assessment.

Toxicological Evaluation of Mixtures of Uncharacterized Substances

Since there are many more compounds present in drinking water sources
than the few that could be considered individually and since the
health concern is from exposure to mixtures, it is therefore necessary
that relevant mixtures be tested and evaluated. The toxicological
assessment of treated wastewater is conceived as an evaluative
process, progressing through three phases of testing. In each phase,
the mixtures from concentrated treated wastewater are compared with
concentrated mixtures from potable water obtained from a conventional
source in order to make a relative or comparative assessment of
potential health hazards and risks. Although the conventional water
may be regarded as "safe" by virtue of its compliance with current
drinking water standards, developments in toxicity testing methodology may demonstrate that the health risks from the conventional
water need to be reassessed.

Phase 1 Toxicity Testing

Phase 1 testing is designed to detect mutagenic and teratogenic
activity and carcinogenic potential, as well as to give an indication
of acute target organ toxicity. This initial comparative assessment
of toxicity results in the characterization of treated wastewater as
having a hazard equal to, greater than, or less than the hazard
potential presented by conventional water. Certain positive results
in this phase of testing would lead to the conclusion that there is
an increased hazard potential for the reused water. In such cases,
further testing would be unnecessary. If both the conventional and
reused water gave reproducibly negative results in this phase, a more
complete evaluation should be conducted in Phase 2 (and possibly
Phase 3). Phase 2 is designed to detect subchronic (90-day) toxicity
and reproductive toxicity. Phase 3 is intended to detect carcinogenicity as well as any other chronic adverse effects. Table 7-1
lists the proposed testing phases and provides analyses of and
comments on potential outcomes.

For *in vitro* tests and *in vivo* cytogenetics assays, duplicate
analyses are needed for all samples. However, because of the nature
of the teratogenicity tests and 14-day, repeated-dose toxicity assays
and the time required to perform them, duplicate analyses on the same
sample are not required.

The *in vitro* assays and *in vivo* cytogenetics analyses should be
performed on water concentrates on a monthly schedule. The teratogenicity and the 14-day, repeated-dose toxicity determinations need
to be performed 3 times per year on three different samples. If,
after 2 or 3 months of testing, reproducibly positive results are
obtained in Phase 1 testing, evaluation of water concentrates need
not proceed to Phases 2 and 3; further testing in these models need
not continue until some alternative is implemented, e.g., alteration
of process treatment.

TABLE 7-1 Toxicity Tests

PHASE 1

Conventional Water	Reused Water

In Vitro Tests

Mutagenicity	Mutagenicity
In vitro transformation	In vitro transformation

In Vivo Tests

Acute toxicity	Acute toxicity
Teratogenicity	Teratogenicity
Short-term, repeated dose studies--	Short-term, repeated dose studies--
14-day (includes in vivo cytogenetic assay)	14-day (includes in vivo cytogenetic assay)

Comparative Analysis

1. Negative results (no detectable toxicity) in all assays on both water sources: Proceed to Phase 2.

2. Positive results[a] in Phase 1 assays on reused water and negative results on conventional water indicate health risks are greater for reused than for conventionally treated water. Options include the following:

   a. Review treatment process and make improvements;
   b. do not use water for human consumption; or
   c. use reused water for human consumption during emergencies only.

3. Positive results[a] on conventional water and negative results on reused water indicate health risks are less for reused water and testing should proceed to Phase 2. The assumption is made that use of conventional water represents an acceptable risk by local and federal authorities.

4. Positive results[a] on both types of water: Quantitatively determine the levels of potency, and ascertain if reused water represents a greater or lesser health risk. If the response is less with reused water and conventional water is believed to represent an acceptable risk, proceed to Phase 2 testing. If health risks are greater for reused water, then options under 2 (above) should be considered.

TABLE 7-1 Continued

5. Positive results in bacterial mutagenicity tests, but negative results in mammalian cell gene mutation, *in vivo* cytogenetics tests, and *in vitro* transformation should be properly noted, but need not stop progression of testing into Phase 2. This analysis recognizes the current inability to draw conclusions about potential health risks based on a single positive result in a prokaryotic system.

6. A positive teratogenicity test indicates that the water contains compounds that are potentially teratogenic in humans. Options include the following:

    a. Review treatment process and make improvements;
    b. do not use the water for human consumption;
    c. determine an estimate of risk by comparing the reused water with conventional water, which is assumed to represent an acceptable risk; or
    d. use reused water for human consumption during emergencies only.

Comments

1. The effect that any single positive result or group of positive results will have on progression through the several phases of testing should be determined by individuals qualified in the appropriate toxicological sciences.

2. Positive results showing either genotoxicity or transformation activity indicate that the water contains compounds that have potential genotoxic or carcinogenic risk for humans.

3. Positive results[a] in the 14-day, repeated-dose study indicate the water is potentially toxic in humans. Inconclusive results in that study may have to be resolved by a longer period of testing such as the subchronic 90-day Phase 2 study. Options include the following:

    a. Review treatment process and effect improvements, or
    b. proceed to Phase 2.

TABLE 7-1 Continued

PHASE 2

Conventional Water	Reused Water
1. Subchronic 90-day study in at least one rodent species, preferably in two species	1. Subchronic 90-day study in at least one rodent species, preferably in two species
2. Reproductive toxicity	2. Reproductive toxicity

Comparative Analysis

1. Negative results (no detectable toxicity) in both water sources: Proceed to Phase 3.

2. Positive results[a] on reused water concentrates; no detectable toxicity or less severe effects with concentrates from conventional water: Reused water represents a greater health risk than does conventional water. Options include the following:

   a. Review treatment process and effect improvements, if possible;
   b. do not use water for human consumption; or
   c. use reused water for human consumption during emergencies only.

3. Positive results[a] on conventional water and negative results on reused water: Health risks are less for reused water and testing should proceed to Phase 3. The assumption is made that use of conventional water represents an acceptable risk by local and federal authorities.

4. Positive results[a] on both types of water: Quantitatively determine the levels of potency and ascertain if reused water represents a greater or lesser health risk. If the response is less with reused water and conventional water is believed to represent an acceptable risk, proceed to Phase 3 testing. If health risks are greater for reused water, then options under 2 (above) should be considered.

TABLE 7-1 Continued

PHASE 3

Conventional Water | Reused Water

1. Chronic lifetime feeding study in one species of rodent

1. Chronic lifetime feeding study in one species of rodent

Comparative Analysis

1. Positive results[a] on reused water and negative results on conventional water indicate health risks are greater for reused than for conventionally treated water. Options include the following:

    a. Review treatment process and make improvements;
    b. do not use water for human consumptions; or
    c. use reused water for human consumption during emergencies only.

2. Positive results[a] on conventional water and negative results on reused water indicate health risk are less for reused water. The assumption is made that the use of conventional water represents an acceptable risk by local and federal authorities.

3. Negative results in both sources indicate that the risk is no different within limitations of methods.

---

[a]Positive results should be interpreted by individuals with expertise in the various aspects of toxicology represented by the tests. The following factors must be considered when positive results are obtained:

- types and locations of lesions and their differential importance
- severity of lesions in same organs
- response potency (number of animals responding/dose)

Evaluations of positive or negative results in any individual assay are dependent on the demonstration of dose-response relationships, statistically significant increases over those of controls, or both. Comparisons between conventional and reused water must also be made on the basis of statistically significant and reproducible differences derived from test results on equal weight of total solutes.

Phases 2 and 3 Toxicity Testing

These two phases of evaluation deal with cumulative toxicity. Phase 2 consists of a subchronic, 90-day study in a rodent species (preferably in two species, one a nonrodent) and includes an assessment of reproductive toxicity. Phase 3 is a combination chronic toxicity (including carcinogenicity) study. These two phases of testing are especially important for substances for which there is long-term repeated exposure. An inherent difficulty in these evaluations, as well as in other phases of testing reused water, is a consequence of using concentrates. Caution should be exercised in interpreting these data to ensure that the effects observed are not a phenomenon of the high doses necessary to elicit a response. These mixtures may produce effects that would not be realized if such solutes were present in an unconcentrated form. Recognition of this possibility makes it all the more important that qualified toxicologists evaluate results of all three phases of testing. Dose-response relationships and types of lesions that can result from testing concentrated mixtures need careful interpretation in order to reach reasoned judgments on potential health effects for humans. Depending on geographic location and consequent solute composition, consumption by humans may not entail undue increased risk, whereas other geographic areas may introduce contaminants that pose greater risk to human health.

Monitoring and Testing in Actual Reuse

For the most part, the extensive monitoring and toxicological testing described in this chapter are recommended primarily for the pilot-plant studies of systems being considered for municipal wastewater reuse where product water will be used for human consumption and food processing. Because of the high cost of monitoring and toxicological testing, it would be impractical to require these procedures at a frequency in which the costs far exceed the value of the information acquired in wastewater treatment systems regularly used to produce potable water. Nevertheless, to provide a reasonable assurance that undesirable constituents are not present at concentrations that pose a health risk greater than that of conventionally treated water, it would be advisable to require more extensive monitoring of reused water than is currently specified, e.g., in the Interim Primary Drinking Water Regulations (Environmental Protection Agency, 1975), for conventionally treated public water supplies.

As discussed in Chapter 6, specific chemical and biological constituents as well as surrogate or indicator parameters should be monitored. Subject selection should be related in part to the results of the pilot-plant studies, which may reveal some constituents whose concentrations may be judged to pose unacceptable risks to human health. Such constituents should be considered for routine monitoring. Also, some of the nonspecific chemical parameters (Chapter 6, Table 6-3) should be selected, especially if they can be consistently correlated in the pilot-plant studies with the presence of high concentrations of specific chemicals or higher-than-normal hazards as indicated by the toxicological tests.

Similar considerations should apply to biological pathogens and indicator organisms. At this stage of limited experience with treated wastewater, there is no clearly discernible set of parameters that can be specified for the routine monitoring or MCL's for potable reuse supplies. As understanding of these systems develops and also as the scientific and technical communities become better able to assess the adverse health effects of trace concentrations of specific chemicals or mixtures, it may become possible to specify a priori more precise monitoring requirements for treated wastewater generically, rather than to depend solely on judgments related to the outcome of pilot plant studies at specific sites.

Finally, in those communities where reuse is practical, consideration should be given to an occasional update of the full-scale analytical and toxicological evaluations as recommended for the pilot-plant studies. There are two principal reasons why these evaluations should be made. First, until we have more experience with such testing and evaluation systems, it will be valuable to confirm that there are no long-term adverse health effects as measured by the test systems currently in use. Second, it is likely that the science of toxicological testing and the technology of trace chemical measurement will continue to improve. Because the principal concern with treated wastewater is the possible adverse health effects of trace chemicals, any methodologies that can assist in the assessment of their health consequences should be used as they are developed and validated.

The principal focus and ultimate concern of this study is to reduce the possible risks to human health from the use of treated wastewater (both for direct consumption and in processed foods). Thus, an obvious question is the advisability of epidemiological investigations and of monitoring of human health indices for populations using such water. Although some consideration might be given to undertaking such assessments (such as comparision of groups of humans before and after start-up), it is unlikely that any low levels of incremental risk would be identified in these communities if the previously stated criteria for instituting wastewater reuse for human consumption are met.

Summary

The Panel on Quality Criteria for Water Reuse believes that there is a critical need to evaluate reused water chemically, microbiologically and toxicologically. The panel has concluded that the most practical way to make judgments about the potential health hazards of reused water is to compare it with conventional supplies which have risks, if any, that are presumed to be acceptable. The initial comparisons of conventional with reused water should be done on the basis of identifiable individual compounds and microbiological organisms. Drinking Water Standards, Priority Pollutants, other selected "sentinal" chemicals, and additional microbiological indicators would first be determined. The results of these tests would influence the need to proceed with additional testing since a reused water that failed such a comparison would be rejected as not being "as safe as" a generally accepted conventional supply.

Because of the practical impossibility of identifying and testing all of the individual compounds present in reused water, it is ultimately necessary to test mixtures of chemicals. Since the available toxicological tests are relatively insensitive, it is also necessary that the mixtures be concentrated to increase the sensitivity of the tests.

The final comparison between reused and conventional water is based on the outcomes of a series of tiered tests designed to give information on the relative toxicities of the concentrates from the two water supplies. Phase 1 tests include in vitro assessments of mutagenic and carcinogenic potential by means of microbial and mammalian cell mutation and in vivo evaluations of acute and short-term subchronic toxicity, teratogenicity, and clastogenicity. Phase 2 includes a longer term (90-day) subchronic study and a test for reproductive toxicity. Phase 3 is a chronic lifetime feeding study. It is essential that the results of all the tests be validated and evaluated by well-qualified professionals in order to enhance the utility of the comparative nature of this assessment.

Depending on the results of the various comparative test phases a judgment will be reached that reused water is as safe as, more safe than, or less safe than a conventional water supply which is presumed to be safe. The final decision to use treated wastewater for potable purposes or for food processing can only be made after a careful evaluation of potential health effects, treatment reliability, cost, necessity and public acceptance.

REFERENCES

Environmental Protection Agency. 1975. Interim Primary Drinking Water Standards. Fed. Reg. 40:11990-11998.

Environmental Protection Agency. 1977. National Secondary Drinking Water Regulations. Fed. Reg. 42:17143-17147.

McCarty, P.L., J. Kissel, T. Everhart, R. Cooper, and C. Leong. 1980. Mutagenic activity and chemical characterization for the Palo Alto Wastewater Reclamation and Groundwater Injection Facility. Tech. Rep. No. 250. Health Effects Research Laboratory, U.S. Environmental Protection Agency, Cincinnati, Ohio. 65 pp.

National Academy of Sciences. 1977. Drinking Water and Health. Safe Drinking Water Committee, Advisory Center on Toxicology, Assembly of Life Sciences, National Research Council. National Academy of Sciences, Washington, D.C. 939 pp.

National Academy of Sciences. 1980. Drinking Water and Health, Vol. 3. Safe Drinking Water Committee, Board on Toxicology and

Environmental Health Hazards, Assembly of Life Sciences, National Research Council. National Academy Press, Washington, D.C. 415 pp.

National Academy of Sciences. 1982. Drinking Water and Health, Vol. 4. Safe Drinking Water Committee, Board on Toxicology and Environmental Health Hazards, Assembly of Life Sciences, National Research Council. National Academy Press, Washington, D.C. 299 pp.

World Health Organization. 1979. Human Viruses in Water, Wastewater and Soil. Tech. Rep. Ser. No. 639. World Health Organization, Geneva, Switzerland. 50 pp.

Appendix A
Concentration Methodologies for Preparation of Water Concentrates for Toxicity Testing

Chapters 5 and 7 discuss a toxicity testing program that could be undertaken to assess the health significance of wastewater reuse for potable water supplies. The suggested testing program includes a battery of in vivo and in vitro tests to be conducted on concentrates of advanced wastewater treatment plant effluents that are candidates for potable water reuse, with a conventional drinking water supply system to act as a control. Ideally, toxicity testing should be conducted on concentrates of water samples that are statistically ($p < 0.05$) representative of these two water systems.

No single concentration method for biological testing and chemical analysis is adequate for isolating all the organic constituents from the inorganic constituents and water in the sample matrix. This methodological deficiency is particularly important when large volumes of water must be concentrated. Each concentration method has certain advantages and disadvantages, but all have the potential to alter organic constituents and suffer from differences in specificity (i.e., different chemical classes or groups may not be concentrated or isolated to the same degree) (Jolley, 1981). Thus, the procedures recommended in this appendix to prepare concentrates represent practical choices within stated limitations. Two initial criteria are (1) it is not expected or planned that volatile organics will be concentrated (but these would be evaluated by chemical analysis), and (2) the salt concentrations in the concentrates should be kept below 1% to minimize possible adverse effects on the biological systems used in the toxicological evaluations.

For the toxicity testing proposed, it is not practical to efficiently isolate volatile organics with a boiling point below 100°C. However, these chemicals may be analyzed by the purge and trap method (Bellar and Lichtenberg, 1974; Environmental Protection Agency, 1979; U.S. Geological Survey, 1981). If desired, volatile organics can be reconstituted just before toxicity testing.

Outlines of the suggested testing protocols described in Chapter 5 are shown in Tables A-1 and A-2 for the in vitro and in vivo systems, respectively. The purpose here is not to set forth a rigid scheme, but rather to indicate the extent and nature of the water requirements and concentration factors that might be required for toxicological testing of wastewater renovation systems. It is apparent that the volumes of water to be concentrated are much smaller for the in vitro as compared to the in vivo systems. In

TABLE A-1 In Vitro Tests

Test	No. of Plates	Doses (mg TOC[a]/ plate)	Total TOC[a] Required (mg)	Volume Required to Concentrate (liters)	Concentration Factor
Ames/Salmonella	36 (4 strains, 3 doses, each in triplicate)	10, 1, 0.1	133	44[b] or 22[c]	To dryness
Mammalian cell	9 (1 cell type, 3 doses, each in triplicate)	1, 0.1, 0.01	10	3.3[b] or 1.7[c]	To dryness
In vitro transformation	15 (1 cell type, 3 doses, 5 plates/dose)	1, 0.1, 0.01	17	6[b] or 3[c]	To dryness

Assuming water to be concentrated contains:
[a] TOC = total organic carbon.
[b] 3 mg TOC/liter (and 300 mg total dissolved solids/liter).
[c] 6 mg TOC/liter (and 600 mg total dissolved solids/liter).

NOTE: Each test is to be performed monthly and separately in both the renovated water and the conventional water. The listed requirements are those of each month's tests for each type of water being tested.

TABLE A-2 In Vivo Tests

Test	Frequency	Exposure Period	Number of Animals/Dose	Number of Dose Levels	Concentration Factor	Volume of Water to be Concentrated/Exposure Period/Test (liters)[c]
Teratogenicity with two species	3/yr	5 days 5 days	Mouse (25) Rat (25)	3 3	400 - 1,700 650 - 2,700	700[b] - 1,400[a] 9,300[b] - 19,000[a]
Toxicity (acute)	1/yr	1 day	Rat (10: 5M, 5F)	3	100, 500, 1,000	400
Short-term repeated dose (14-day)	3/yr	14 days	Rat (20: 10M, 10F)	3	100, 200, 400	4,900
Subchronic	1/yr	75 days	Rat (40: 20M, 20F)	3	100, 200, 400	52,500
Reproductive phase, F_{1a} (follows subchronic test)	1/yr	75 days	Rat (40: 20M, 20F)	3	100, 200, 400	52,500
Reproductive phase, F_{2a}; (follows reproductive phase F_{1a})	1/yr	75 days	Rat (40: 20M, 20F)	3	100, 200, 400	52,500
Chronic lifetime	Once	24-30 mo	Rat (100: 50M, 50F)	3	100, 200, 400	$1.3 - 1.6 \times 10^6$

Assumes that the unconcentrated water contains:
[a] 3 mg TOC/liter.
[b] 6 mg TOC/liter.
[c] Based on a water intake of 3 ml/day/mouse; 25 ml/day/rat.

NOTE: Each test is to be performed separately on both the renovated water and conventional water. The listed requirements are those for each type of water and each test; the frequency is also indicated.

fact, the latter require substantial amounts of water. For the chronic lifetime study alone, 1,750 liters of water per day would have to be concentrated down to a volume of 2.5 liters to provide test dosages for administration to 100 rats <u>for each water type being tested</u>. Within this requirement, three different aliquots of the 1,750 liters would require separate concentration factors of 100, 200, and 400, respectively, as indicated in Table A-2. It is thus apparent that preparation of the water concentrates for toxicity testing requires both a substantial and complex effort. Also, if archival samples and/or samples for chemical analysis are needed or required, then the water sample size must be increased accordingly.

Before beginning the toxicity testing program, each of the concentration procedures recommended must first be tested at the water treatment site (sample collection site) to ensure adequacy of the concentration method, e.g., solubility of the components, minimization of artifacts, and development of a quality assurance program. A useful approach would be to perform a mass balance, based on total and purgeable organic carbon during this initial testing period.

RECOMMENDED CONCENTRATION PROCEDURES

Sample Volumes < 100 Liters, High Concentration Factor

Lyophilization (freeze drying) is a feasible process for concentrating limited numbers of 50- to 100-liter samples to relatively high degrees of concentration (e.g., 3,000-fold to dryness). Thus, lyophilization is one method of choice for preparing samples for the <u>in vivo</u> tests.

Bacterial mutagenesis tests have been conducted using distilled water solutions of the freeze-dried residues (wastewater effluents concentrated up to 3,000-fold) (Cumming et al., 1979) and partially freeze-dried samples (wastewater effluents concentrated 10-fold) (Foster and Wilson, 1981). High salt concentrations in such concentrates may cause toxicity problems in the bacterial tests. The use of dimethylsulfoxide (DMSO) or methanol to extract the organic constituents from the freeze-dried residues for mutagenicity tests should be investigated.

Dialysis of concentrated solutions of the freeze-dried residues to remove the high salt concentration can result in unacceptable losses of low molecular weight (MW) organic compounds. Ultrafiltration with 1,000-MW cutoff membranes will remove essentially all inorganic salts (F. Lightly, Osmonics, Inc., personal communication, 1982), but will result in unacceptable loss of low MW organic compounds. The 200-MW cutoff ultrafiltration membranes reject only 5% of the inorganic salts; consequently, the concentrates would have to be diluted more than 20-fold with distilled water during the ultrafiltration process (diafiltration) to desalt the concentrate. This use of large volumes of distilled or deionized water to "rinse" the concentrate may be unacceptable because of possible introduction of artifacts or toxic materials (Cheh et al., 1981).

A second method of choice is adsorption on solid adsorbents. Adsorption on XAD resins (Junk et al., 1976) and extraction of the adsorbed organics with methanol or acetone (Flanagan and Allen, 1981; McCarty et al., 1980; Neal et al., 1980) may be used conveniently with 50- to 100-liter sample volumes. In addition to such solvents, some investigators use dilute acid and base washes to facilitate removal of adsorbed organics from XAD resins, followed by neutralization of the washes and ether extraction (Kopperman et al., 1978). However, recovery of organic constituents from water samples by XAD adsorption is limited (e.g., only 5%-20% of the total organic carbon is recovered, although this recovery included 60%-80% of the neutral organic compounds) (Foster and Wilson, 1981).

The resin bed used should have sufficient capacity for the sample to be concentrated. If capacity is not carefully determined for a particular source water, the sample collected will contain only the well-adsorbed species because of competitive adsorption (Suffet et al., 1982). The dechlorination of a sample by sulfite before the application of the water to a resin column has been shown to decrease the mutagenic activity of a sample (Cheh et al., 1980). Therefore, the use of a dechlorination agent before applying the water to a resin bed must be evaluated against the artifact formation in a resin bed from the reaction of chlorine with the resin matrix.

Use of solvent extraction is also a potentially feasible process. This is an engineering unit operation that is well-adapted to continuous processing. It has been successfully used to isolate nonpolar compounds of boiling point >100°C (Yohe et al., 1979). Solvent extraction may be a useful process for routinely concentrating 50 to 100 liters of water. Its major drawback is evaporation and recovery for reuse of large volumes of the organic solvent. Other problem areas that must be considered are purification of sufficient solvent and minimization of artifact formation by heat.

Sample Volumes > 100 Liters, Medium Concentration Factor

Long-term daily concentration of large-volume samples (e.g., for lifetime in vivo tests) requires that the concentration procedure be reliable, relatively easy to maintain, and capable of being operated continuously for long periods. Thus, because of proven utility in industrial applications, ultrafiltration is one method of choice for concentrating large-volume samples, particularly if the toxicity tests can be conducted on concentrates of <1,000 MW-organic constituents. Ultrafiltration with 1,000-MW cutoff membranes and essentially complete salt rejection can be used rouitinely to process large volumes of water for preparation of essentially "salt-free" concentrates (F. Lightly, Osmonics, Inc., personal communication, 1982).

Use of ultrafiltration membranes with increasingly lower MW cutoffs produces concentrates with increasingly higher salt concentrates. For example, 200-MW cutoff membranes reject only 5% salt. Consequently, the concentrates must be diluted with distilled water and ultrafiltered again. The diafiltration process is conducted continuously, but may require large-scale dilution or rinsing of the concentrate. However, use of very large volumes of dilution water may be unacceptable.

Reverse osmosis (RO) is essentially the same process as ultrafiltration with limited salt rejection. Kopfler and coworkers (1977) used RO in combination with solvent extraction and XAD adsorption. Solvent extraction with pentane and methylene chloride was used to remove organics from the RO retentate, i.e., for desalting, and XAD was used for adsorption of the intractables from the extracted retentate. Although limited in recovery of organics (30% to 40%), this combination procedure has been used successfully to concentrate large volumes of water.

Use of solid adsorbents (e.g., XAD) and solvent extraction represent feasible engineering processes for concentrating large volumes of water. However, scaleup, cleaning, and preparing large quantities of solid adsorbent may be technically difficult to accomplish. As an engineering process, solvent extraction should be readily applicable to continuous processing of large-volume samples. However, the caveats mentioned previously are applicable, particularly to large-volume processing. More work is needed in this area.

GENERAL CONSIDERATIONS

Sampling

Composite or continuous sampling provides water samples more representative of renovated water effluents. If possible, such samples should be processed immediately to avoid changes during storage. However, collection of small-volume samples may be more easily obtained by batch sampling.

Storage

If samples must be stored for short periods before concentration, they should be kept just above 0°C. Concentrates or dried organic residues from extraction or lyophilization processes should be stored at -40°C or lower. There is little information regarding the stability of samples stored cryogenically. Long-term storage of large samples should be avoided because of the considerable length of time required to thaw samples of 50 to 100 liters. Thus, research is needed to determine the best method for storage of such samples to prevent their degradation and the development of artifacts.

SUMMARY

A variety of techniques can be used to concentrate water for toxicity testing; all have some limitations. Thus, a combination of procedures is recommended to improve the recoverability of the wide range of compounds likely to be encountered. For example, if the sample volumes are less than 100 liters and if high concentrations are required, lyophilization and ultrafiltration can be used. In contrast, if sample volumes are greater than 100 liters and if medium concentration is desired, ultrafiltration followed by dialysis might be appropriate.

Reverse osmosis is another alternative. In any event, the procedures or their combination should be evaluated for the specific water being tested in terms of recoverability of either specific organic constituents or gross organic parameters such as total organic carbon.

REFERENCES

Bellar, T.A., and J.J. Lichtenberg. 1974. Determination of volatile organics at micrograms per liter levels by gas chromatography. J. Am. Water Works Assoc. 66:739.

Cheh, A.H., J. Stochdopole, C. Heilig, P.M. Koski, and L. Cole. 1980. Destruction of direct-acting mutagens in drinking water by nucleophiles: Implications for mutagen identification and mutagen elimination from drinking water. Pp. 803-815 in R.L. Jolley, W.A. Brungs, and R.B. Cumming, eds. Water Chlorination: Environmental Impact and Health Effects, Vol. 3. Ann Arbor Science, Ann Arbor, Mich.

Cheh, A.M., R.E. Carlson, J. Hilderbrandt, C. Woodward, and M. Pereira. 1981. Contamination of purified water by mutagenic electrophiles. Presented at the Fourth Conference on Water Chlorination: Environmental Impact and Health Effects, Pacific Grove, Calif., October 18-23.

Cumming, R.B., L.R. Lewis, R.L. Jolley, and C.I. Mashni. 1979. Mutagenic activity of nonvolatile organics derived from treated and untreated wastewater effluents. In A.D. Venosa, ed. Progress in Wastewater Technology. EPA 600/9-79-018. U.S. Environrmental Protection Agency, Cincinnati, Ohio.

Environmental Protection Agency. 1979. Guidelines for establishing test procedures for the analysis of pollutants, proposed regulations. Fed. Reg. 44 (233):G9463-9575.

Flanagan, E.P., and H.E. Allen. 1981. Effects of water treatement on mutagenic potential. Bull. Environ. Contam. Toxicol. 27:765-772.

Foster, R., and I. Wilson. 1981. The application of mutagenicity testing to drinking water. J. Inst. Water Eng. Sci. 35(3):259-274.

Jolley, R.L. 1981. Concentrating organics in water for biological testing. Environ. Sci. Technol. 15(8):874-880.

Junk, G.A., J.J. Richard, J.S. Fritz, and H.J. Svec. 1976. Resin sorption methods for monitoring selected contaminants in water. Pp. 135-153 in L.H. Keith, ed. Identification and Analysis of Organic Pollutants in Water. Ann Arbor Science, Ann Arbor, Mich.

Kopfler, F.C., W.E. Coleman, R.G. Tardiff, S.C. Lynch, and K.K. Smith. 1977. Extraction and identification of organic micropollutants: Reverse osmosis method. Ann. N.Y. Acad. Sci. 298:20.

Kopperman, H.L., D.W. Kuehl, and G.E. Glass. 1978. Chlorinated compounds found in waste treatment effluents and their capacity to bioaccumulate. Pp. 311-328 in R.L. Jolley, ed. Water Chlorination: Environmental Impact and Health Effects, Vol. 1. Ann Arbor Science, Ann Arbor, Mich.

McCarty, P.L., J. Kissel, T. Everhart, R. Cooper, and C. Leong. 1980. Mutagenic activity and chemical characterization for the

Palo Alto wastewater reclamation and groundwater injection facility. EPA/1-81-016. Health Effects Research Laboratory, U.S. Environmental Protection Agency, Cincinnati, Ohio. 65 pp.

Neal, M.W., L. Mason, D.J. Schwartz, and J. Saxena. 1980. Assessment of mutagenic potential of mixtures of organic substances in renovated water. EPA/1-81-016. Health Effects Research Laboratory, U.S. Environmental Protection Agency, Cincinnati, Ohio.

Suffet, I.H., B.A. Najar, and J. Gibbs. 1982. Quality assurance of macroreticular resin (MRR) samplers in trace organic analysis. Paper presented at the 183rd American Chemical Society National Meeting, Las Vegas, Nev. March 28-April 2, 1982.

U.S. Geological Survey. 1981. Determination of acetone and methyl ethyl ketone in water. Pp. 78-123 in Water Resources Investigations. PB 291151. U.S. Geological Survey, Washington, D.C.

Yohe, T.L., I.H. Suffet, and R.J. Grochowski. 1979. Development of a teflon helix continuous liquid-liquid extraction apparatus and its application for the analysis of organic pollutants in drinking water. Pp. 47-67 in C.E. Van Hall, ed. Measurement of Organic Pollutants in Water and Wastewater. ASTM STP 686. American Society for Testing and Materials, Philadelphia, Pa.

Appendix B
Further Statistical Details on Sampling

The mean, variance, and coefficient of variation for a composite sample are developed within this section. Given n independent sample values $X_1, X_2, \ldots, X_n$ with expectations $\mu_1, \mu_2, \ldots, \mu_n$ and variances $\sigma_1^2, \sigma_2^2, \ldots, \sigma_n^2$, the expectation and variance of the composite of $X_1, X_2, \ldots, X_n$ represented by

$$Z_n = \sum_{i=1}^{n} a_i X_i,$$

where $\sum_{i=1}^{n} a_i = 1$, can be calculated as

$$E(Z_n) = E\left(\sum_{i=1}^{n} a_i X_i\right)$$

$$= E\left[E\left(\sum_{i=1}^{n} a_i X_i \,\middle|\, \underset{\sim}{a}\right)\right]$$

$$= E\left(\sum_{i=1}^{n} a_i \mu_i\right)$$

$$= \sum_{i=1}^{n} E(a_i)\, \mu_i$$

and

$$\text{var}(Z_n) = E\left[\text{var}\left(\sum_{i=1}^{n} a_i X_i \,\middle|\, \underset{\sim}{a}\right)\right] + \text{var}\left[E\left(\sum_{i=1}^{n} a_i X_i \,\middle|\, \underset{\sim}{a}\right)\right]$$

$$= E\left(\sum_{i=1}^{n} a_i^2 \sigma_i^2\right) + \text{var}\left(\sum_{i=1}^{n} a_i \mu_i\right)$$

$$= \sum_{i=1}^{n} (\sigma_a^2 + \mu_a^2)\sigma_i^2 + \sum_{i=1}^{n}\sum_{i=j}^{n} \text{cov}(a_i, a_j)\, \mu_i \mu_j,$$

where $\underset{\sim}{a} = (a_1, a_2, \ldots, a_n)$ and μ_a, σ_a^2, and $\text{cov}(a_i, a_j)$ are the mean, variance, and covariances of the a's, respectively.

If compositing is perfect ($a_1 = a_2 = \ldots = a_n = \frac{1}{n}$), then

$$E(Z_n) = \frac{1}{n}\sum_{i=1}^{n}\mu_i = \bar{\mu}_n \text{ and } \text{var}(Z_n) = \frac{1}{n^2}\sum_{i=1}^{n}\sigma_i^2.$$

Then, the coefficient of variation (CV) for the composite is

$$CV(Z_n) = \frac{(\frac{1}{n^2}\sum_{i=1}^{n}\sigma_i^2)^{\frac{1}{2}}}{\sum_{i=1}^{n}\frac{\mu_i}{n}} = \frac{(\sum_{i=1}^{n}\sigma_i^2)^{\frac{1}{2}}}{\sum_{i=1}^{n}\mu_i}.$$

If $\sigma_1^2 = \sigma_2^2 = \ldots = \sigma_n^2 = \sigma^2$, then,

$$CV(Z_n) = \frac{\sqrt{n}\,\sigma}{\sum_{i=1}^{n}\mu_i} = \frac{\sigma}{\sqrt{n}\,\bar{\mu}_n},$$

so that the coefficient of variation for a composite is "reduced" by a factor of $\frac{1}{\sqrt{n}}$. This seems to be in keeping with the results from Water Factory 21.

It has frequently been noted that measurements of water constituents are described by a log-normal distribution. That is, $y_i = \ln x_i$ are normally distributed. Then the confidence interval for the average of the logarithms is

$$\bar{y} \pm \frac{ts_{\ln x}}{\sqrt{n}},$$

where $s_{\ln x} \simeq s/x = c$, the coefficient of variation. Confidence limits on the untransformed scale are obtained by taking the anti-logarithms. This results in the confidence limits being determined within a factor $F = \exp(tc/\sqrt{n})$ of the mean. The values of t for 95% confidence limits range from 2.78 for 5 samples down to 1.96 for an infinite number of samples. If we use t = 3 to cover the case for a small number of samples, the mean generally would be estimated within a factor of $F = \exp(3c/\sqrt{n})$. Solving for n gives an estimate of the number of samples required to estimate the mean within a factor of F,

$$n = \left(\frac{3c}{\ln F}\right)^2.$$

For example, suppose the coefficient of variation were 0.5 and one desires to estimate the mean within a factor of 2, then n = $(3 \times 0.5/\ln 2)^2$ = 4.68 or 5 samples would be required. If the calculated value of n were much different than 5, then the value of 3 should be replaced by the appropriate value of t and n recalculated.

Water constituents are monitored by sampling. Water samples may be a grab sample at some point, a composite of several grab samples over a period of time, or a continuous sample over a period of time. The question being addressed is how representative is a sample over some longer period? Let x_i represent the measurement of a water constituent from the ith sample, where i = 1,2, ..., n samples. If the x_i's are identically independent, normally distributed measurements, then the confidence limits for the true average value of the constituent is

$$\bar{x} \pm \frac{ts}{\sqrt{n}},$$

where x is the mean value of the x_i's, the value is selected for the desired level of confidence with (n - 1) degrees of freedom, and the standard deviation is given by

$$s = \sqrt{\frac{\Sigma (x_i - \bar{x})^2}{n - 1}}.$$

It is desirable to restrict the width of the confidence interval to ±w; then the expected number of samples required is

$$n = (ts/w)^2.$$